the JOBS TO BE DONE *playbook*

Align Your Markets, Organizations, and Strategy Around Customer Needs

by

JIM KALBACH

foreword by Michael Schrage

TWO WAVES BOOKS
NEW YORK, NEW YORK, USA

The Jobs to Be Done Playbook

Align Your Markets, Organization, and Strategy Around Customer Needs

By Jim Kalbach

Two Waves Books
an imprint of Rosenfeld Media
125 Maiden Lane, Suite 209
New York, New York
10038 USA

On the Web: twowavesbooks.com
Please send errors to: errata@twowavesbooks.com

Publisher: Louis Rosenfeld
Managing Editor: Marta Justak
Illustrations: Michael Tanamachi
Interior Layout Tech: Danielle Foster
Cover and Interior Design: The Heads of State
Indexer: Marilyn Augst
Proofreader: Sue Boshers

ISBN: 1-933820-68-3
ISBN 13: 978-1-933820-68-2
LCCN: 2019948483

Printed and bound in the United States of America

For my brother

Contents at a Glance

Contents and Executive Summary

Foreword ix

Introduction xi

At a time when consumers have unprecedented choices, JTBD offers a way of seeing markets to maximize growth around customer needs. The core message of JTBD is clear: focus on people's objectives, not on your company, offering, or brand.

Chapter 1: Understanding Jobs to Be Done 1

JTBD is a broad field with a decades-old history and varying perspectives unified by a set of shared principles. Begin your JTBD by understanding the core tenets of the theory and practice.

Chapter 2: Core Concepts of JTBD 15

A common language of JTBD helps filter otherwise irregular market feedback into a normalized system. Apply the JTBD framework to better predict your customer-driven growth.

Chapter 3: Discovering Value 47

People value products and services that help them get a job done. Find your opportunities by interviewing people in your market and mapping their job.

Chapter 4: Defining Value 83

JTBD provides a way to model various aspects of your market. Identify unmet needs and define your target markets all through the JTBD lens.

Chapter 5: Designing Value 119

JTBD is versatile and provides many techniques for developing solutions. After understanding the problem you are solving, use JTBD to design products and services that customers really want.

Chapter 6: Delivering Value 161

Offering solutions that get a core job done can create a strong market pull, but that's often not enough. Use JTBD to accelerate your go-to-market efforts as well.

Chapter 7: (Re)Developing Value 195

JTBD not only frames innovation and marketing efforts, but it also provides a way of grounding strategy in real-world observations. Use various techniques to guide your business at the highest levels.

Chapter 8: JTBD in Action 223

The practice of JTBD includes various methods and techniques. Find the best combination of plays to fit your situation and address your specific challenges.

Final Thoughts 253

Quick Reference: JTBD Plays 257

Resources on JTBD 279

Index 287

Acknowledgments 298

About the Author 300

Foreword

Some books are written primarily to be read; others are written mainly to be used. Jim Kalbach's book, *The Jobs to Be Done Playbook*, should inspire the best of both. The reason is not just that Jim is a fluent writer with a crisp and clear purpose, but that he treats the fundamentals of user design, user experience, and the "job to be done" with thoughtfulness, seriousness, and rigor. I cannot overstate how important that is.

In my classes, workshops, and advisory work, I have the good fortune to work with talented people who truly want to do excellent work. I like and admire them. They are talented *and* smart. But oftentimes, because they are so talented and smart, they presume or assume they *know* the fundamentals of something when, in truth, they do not. With apologies to Atul Gawande, they have a "Checklist Manifesto" grasp of what they're trying to do. That is, they're doing everything they're *supposed* to do but lack an essence and esprit that makes the work compelling. What people minimize (or overlook) about Gawande's checklists is that they are supposed to be prompts and reminders for people who truly *know* their stuff. The challenge here is that people—smart people, caring people—don't always quite know the right stuff.

Yes, they know the "product" and the "service" and the desired and desirable "user experience," but do they really understand and appreciate the power and importance of "the job to be done"? The concept is simple and straightforward—it's scalable, implementable, and extensible. Instantiation is not.

That's why Jim's book is so useful and important. He's got the rigor and the chops to not only make the fundamentals accessible and understandable, but practical and doable as well. While I am a huge

Clay Christensen fan (indeed, he was kind enough to write a blurb for one of my books), I think he'd be one of the first to acknowledge that his breakthrough conceptual thinking requires facilitators, translators, and interlocutors to make it work in the real world. With tongue firmly in cheek, it's quite a "job to be done" to get the "job to be done" done. But that's what Jim's book empowers you to do.

This is not a book to be read in a sitting or a transcontinental flight. Similarly, you're a jerk if you hand it to a colleague or a boss without spending some time with it yourself. The real way to get value from this book is to ask yourself—honestly and openly—where your greatest frustration lies as a value creator. Then start leafing through this book—not to find yourself *or* the answer, but to understand the essential fundamentals of the job to be done.

—Michael Schrage
Research Fellow at MIT's Initiative on the Digital Economy and author of *Who Do You Want Your Customers to Become?* (HBR Press)

Introduction

In April 2013, I gave a workshop at UX London, a premier design conference that attracts participants from around the world. As a speaker, I also got to attend the conference. The lineup of presenters was stellar, as it typically is, and I didn't want to miss out on the world-class content presented there. So prior to the event, I scoured the program to see which sessions would be most interesting.

One workshop in particular caught my eye: Des Traynor's "Where UX Meets Business Strategy." This was a three-hour session focused on how design influences business and vice versa. More specifically, Des was going to look into "How to orient a company around a job-to-be-done," as he wrote in the session description.

It couldn't have been a better mix of topics to match my interests at the time—the intersection of design, strategy, and jobs to be done (JTBD). I remember distinctly how intent I was on making it to that session. I had been learning all I could about JTBD and trying some techniques in my work. This was right up my alley. Like Des, I believe that business success comes from understanding human needs and motivations.

I flew to UX London from Hamburg, Germany, where I was living at the time, the day before the event. The organizers had arranged for a car to pick me up from the airport. After I found the driver, he informed me that we'd have to wait a few minutes for another passenger. He pulled out a piece of paper with the person's name on it before standing at the end of the arrivals area. It read "Des Traynor, UX London." I broke into a huge grin.

On the drive from Heathrow to the conference hotel, Des and I talked a lot about jobs to be done. He has a design background and was able to relate to my perspective of the field quite well. We were seeing eye to eye. In fact, during his workshop, Des asked me to speak to the audience of about 80 people for a few minutes, based on my experience with JTBD research.

At its core, the concept of JTBD is straightforward: focus on people's objectives independent of the means used to accomplish them. Through this lens, JTBD offers a structured way of understanding customer needs, helping to predict better how customers might act in the future. The framework provides a common unit of analysis for teams to focus on—the job to be done—and then offers a shared language for the whole team to understand value as perceived from the customer perspective.

Des knew that with decades of history, JTBD could greatly help organizations shift their mindset from inside-out to outside-in. Beyond informing product design, JTBD also has a broad appeal, including marketing, sales, customer success, support, and business strategy.

Des is one of four co-founders of a company called *Intercom*, an online messaging solution that allows businesses to connect directly with customers by coordinating multiple channels of communication. It turns out that JTBD has played a large role in how Intercom approached doing business.

Des and his cofounders innately understood the power of integrating a customer-centric perspective into all aspects of their business. They applied a JTBD mindset to build their company. Writing in the introduction to their recent ebook, *Intercom on Jobs-to-be-Done*, Des reflected on the significance that JTBD had in the company's formation and success:[1]

> When we were first introduced to Jobs-to-be-Done, it quickly resonated with something we already intuitively knew—that great products were built around solving problems. What Jobs-to-be-Done gave us was a better way of framing what we felt—a vocabulary and framework to

1. See *Intercom on Jobs-to-be-Done*: https://www.intercom.com/books/jobs-to-be-done

> unite the team behind a product strategy. Over time, it turns out it's not just a great way for thinking about product. It's become a marketing strategy at Intercom, as well as informing research, sales, and support.

JTBD permeated how Intercom did business, aligning employees across the company. For instance, when UX researcher Sian Townsend left Google to join the company in 2014, she hadn't heard of JTBD. But two years later, she was not only convinced that the framework could help a company be successful, but she also became a dedicated convert. In her talk, "Jobs to Be Done: From Doubter to Believer," Sian highlighted the significance of JTBD at Intercom:[2]

> We've used JTBD to bring great focus to our company. And in the course of using JTBD, we've actually raised a huge amount of money over the last two years. So I feel like we must have been doing something right. It certainly *feels* like it's helped us.

And after years of rapid growth, Intercom received $125 million US dollars in funding in 2018. The company trajectory continues upward. JTBD is a clear part of their success story.

But JTBD doesn't just help startups or small teams align around customer needs. Even large organizations have relied on JTBD for strategic guidance. Just consider Intuit, the tax software giant. After 25 years of existence, the company continues double-digit growth in spite of the fact that the average lifespan of S&P 500 companies is now under 20 years.

So why isn't Intuit dead? For one, Intuit also makes courageous moves, expanding into new markets quicker than others, often through acquisition (e.g., Mint.com and Quicken Loans). And employees are also encouraged to take risks. Experimentation is part of the company culture.

2. See Sian Townsend's presentation "Jobs to Be Done: From Doubter to Believer" given at Front 2016 in Salt Lake City: https://www.intercom.com/blog/videos/jobs-to-be-done-doubter-believer/

But Intuit doesn't just make strategic guesses. Underpinning its seemingly leap-of-faith decisions is a firm grounding in customer needs at the executive level. Focusing on the job to be done allows Intuit to find continued opportunity for growth and consistently provide solutions that customers value. As founder and chairman Scott Cook says, "Jobs Theory has had—and will continue to have—a profound influence on Intuit's approach to innovation."[3]

More and more, there's a growing belief that customer-centric businesses perform better than traditional organizations. For example, a 2014 study by Deloitte showed that customer-centric companies were more profitable.[4] JTBD offers a consistent way to innovate around customer needs across the company, regardless of department or function. This book will show you how to implement JTBD.

A Way of Seeing

The context of business has changed. Consumers have real power: they can research your company's background, compare customer ratings, and find better alternatives all with a simple tap. Traditional approaches to sustaining an enterprise no longer suffice. Operational efficiency, while important, isn't enough to survive.

Within this new business landscape, opportunities for growth come from the outside, from beyond the borders of an organization. But despite a clear customer-centric imperative, traditional businesses have largely failed to change how they think about providing value. For one, they are stuck in management theories of the past, favoring unsustainable goals centered on maximizing short-term profit. But they also believe they can create solutions that customers will truly value without their input.

3. Quoted in a testimonial for Clayton M. Christensen et al., *Competing Against Luck* (New York: HarperBusiness, 2016).
4. See the white paper: "Customer-Centricity: Embedding It into Your Organisation's DNA," https://www2.deloitte.com/content/dam/Deloitte/ie/Documents/Strategy/2014_customer_centricity_deloitte_ireland.pdf

Part of the challenge is that people's decisions and actions are seemingly unpredictable. Resting your growth strategy on fuzzy concepts like "needs" and "empathy" is daunting. While psychology and other fields have precise definitions of human needs, business does not. As a result, risk-averse organizations struggle to grasp the customer perspective and align to it.

JTBD helps shift a collective mindset, from focusing on the organization and its solution to focusing on customers and their needs. More than a particular method, JTBD offers a *way of seeing* your markets, your organization, and your strategy. It's a way of reframing problems and solutions that I refer to as "JTBD thinking" or "jobs thinking" throughout the book.

More importantly, JTBD can help achieve alignment of teams in an organization. The common unit of analysis—the job to be done—provides not only a lingua franca across disciplines, but also fosters a new type of collaboration necessary for today's fast-paced business world.

In my last book, *Mapping Experiences*, I talked at length about aligning to the customer experience. In a nutshell, an experience map and related techniques are devices to get that kind of external alignment.

But in order for teams to have their work aligned, they need a shared focal point. JTBD is no silver bullet, but it provides a starting point for tying alignment to the customer with alignment across teams and departments. This approach goes a long way toward speeding up decision-making and reducing coordinate costs.

What's more, research shows that organizations that involve more of the company are more profitable.[5] In modern companies, culture and collaboration play an increasingly key role. Jobs thinking informs a broad culture of innovation with a common language and perspective.

5. Dylan Minor, Paul Brook, and Josh Bernoff, "Are Innovative Companies More Profitable?" *MIT Sloan Management Review* (blog), December 28, 2017, https://sloanreview.mit.edu/article/are-innovative-companies-more-profitable/

About This Book

This book is your introduction and guide to the nearly thirty-year-old field of JTBD. It is a collection of existing practices that I call "plays," borrowing from the metaphor of a sports playbook. Basically, plays are individual techniques that form the building blocks of work in JTBD. You can use them individually or combine them with other plays, which I illustrate in the last chapter. Each play is indicated with the designation "**PLAY ➤**".

The intent is to give you a practical reference to various approaches within the JTBD canon. But keep in mind that this collection is not comprehensive. I include references for you to learn more on the topic. My aim is to open up a broader discussion and exploration of the role of JTBD in business.

The material included here is based on my own investigation into JTBD and various uses of techniques in my work over the past decade. My goal isn't to introduce a new, competing view of JTBD, nor to present an original method. Instead, I strive to tie existing approaches together by recasting JTBD as a perspective or lens for how to see customers. My hope is that by exposing you to a range of possibilities, you'll be inspired to adopt JTBD thinking and practices into your own work.

In writing this book, I've become acutely aware of the shortcomings of JTBD approaches. Like any method, there are trade-offs. Know that JTBD is certainly no panacea. As you read this book critically, I also encourage you to keep the benefits of JTBD in mind as well.

- First, having a clear unit of analysis—the job—provides a tangible focal point. Needs, emotions, and aspirations are then seen in relation to the job, layered on only after an understanding of the main job.
- Second, because JTBD doesn't originate from one particular discipline, such as design or marketing, other teams can readily adopt the approach and form insights of their own.

- Third, because JTBD views an individual's objective independent of technology, the technique future-proofs your thinking. Framing the job as universally as possible better prepares you to create solutions around how customers may act in the future, not tied to the past.

Working with JTBD may feel uncomfortable at first. It takes practice. The rules of formulating a job statement, for instance, are precise. Start small and experiment, and try not to overcomplicate things: a job is often more straightforward than you think. Eventually, you can have everyone in your organization adopt the lens of JTBD. You'll then have a consistent engine that drives innovation and growth at all levels of an organization.

You might also feel like JTBD overlaps with other existing techniques. In many cases, you'd be right. Just remember that it's unlikely that you will replace existing processes with a JTBD approach all at once. Instead, you'll likely start by introducing parts of JTBD into your current workflow little by little. This book is designed to help you do just that.

The good news is that JTBD is compatible with other modern techniques, such as Design Thinking, Agile, and Lean. Together, these approaches can help you transform the way you do business from end-to-end. As you try the techniques presented here, consider how they may fit into a broader program of modern customer-centered activities.

For instance, people familiar with Intuit will point out after reading the story at the beginning of this introduction that the company also practices user-centered design. Their "Design for Delight" program (aka D4D) actively guides product design and development across the company. So while JTBD is not an instant cure-all, it works well in conjunction with other disciplines and fields.

Finally, keep in mind that techniques in JTBD are still evolving. I continually come across novel uses of jobs thinking, and I encourage you to develop your own approach as you explore the topic. Think of this book as a starting point, a beginning to your understanding of JTBD.

How This Book Is Organized

To structure this book, I asked myself, *what is the core job any business strives to get done*? Peter Drucker reminds us of a good answer: "There is only one valid definition of business purpose: to create a customer… the business enterprise has two—and only these two—basic functions: marketing and innovation."

In other words, businesses exist to create value—value as perceived by customers in satisfying needs (innovation) and value for the company by staying profitable (go-to-market).

From this perspective, I organized the techniques, or plays, in this book around the five stages in providing solutions that customers find valuable:

- **Discover value:** Find the right problem to solve for the people you serve.
- **Define value:** Set the direction for addressing the problem you've identified.
- **Design value:** Create solutions that are desirable, viable, and useful.
- **Deliver value:** Present the solution to the market in a successful business model.
- **(Re)develop value:** Continue to innovate and grow the business.

Figure I.1 illustrates how these stages come together in an iterative motion.

At the center is the core objective: *develop* an offering that people will value. To do that, organizations have to first *discover* and *define* what they believe customers value. This includes primary research and investigation, along with modeling customer behavior and finding the right opportunities. The left side of the diagram represents the innovation imperative that Drucker mentions.

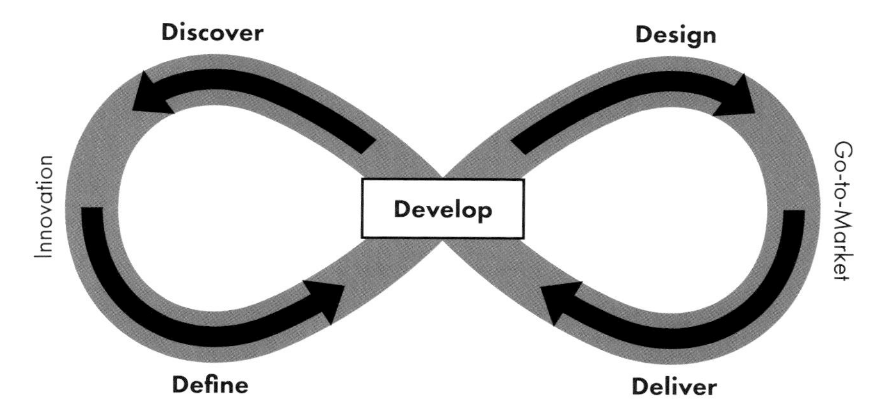

FIGURE I.1 Providing solutions that customers value is an ongoing process between innovation and go-to-market activities.

The right side represents all the activities an organization does to create solutions and introduce them to a market. This begins with the *design* of the offering, which then must be *delivered* to the consumer. It includes everything from conception to planning and marketing to selling.

The cycle repeats: generating business and customer value is about constantly redeveloping your offering. That which is differentiated today will become table stakes tomorrow, and businesses have to constantly reinvent themselves. JTBD provides a consistent way to do that throughout an organization.

Note that this model is not intended to represent a process—rather, it reflects different modes of thinking and operation. A given business will be in all stages at the same time. It's also important to keep in mind that I'm using this model to organize the various plays of JTBD, not as a practical framework of any kind.

After the book introduces JTBD, each of these stages is handled in a separate chapter. I select several plays to illustrate how JTBD can be applied within each mode of thinking. The last chapter brings plays together in methods or "recipes" you can use in different situations.

Who This Book Is For

Most distinctions between strategy and execution are meaningless. Judgment calls happen at all levels of an organization in a cascade of decision-making. In modern knowledge-worker-based companies, strategy isn't confined to the upper echelons only—it happens throughout the enterprise. Managers and individual contributors alike need to be aligned to the same perspective, and everyone contributes to providing solutions that customers will value. JTBD provides a North Star to follow.

The CEO of a company, for instance, could look at customer JTBD to inform strategy. The director of a product line could leverage JTBD to prioritize development. A marketer could use JTBD to help shape messages. Support agents could use JTBD to solve customer issues in a consistent way. There are few parts of an organization that wouldn't benefit from JTBD thinking.

This book is ultimately for change makers and transformation agents inside of companies looking to shift focus toward a customer-centric perspective. It's suited for managers and thought leaders seeking internal alignment around solving customer problems and addressing unmet needs.

More specifically, this book is for people who have limited resources and would like to use JTBD in a lightweight manner. You don't have to hire expensive consultants or execute lengthy, costly projects to benefit from jobs thinking. The techniques presented here reflect a simple view of JTBD so that you can get started right away.

CHAPTER

1

Understanding Jobs to Be Done

IN THIS CHAPTER, YOU WILL LEARN:

- How definitions of JTBD vary
- The origins of JTBD theory and practice
- What divergent schools of thought exist
- The core principles of JTBD

I've done a fair number of field interviews over the years—enough that I don't remember them all. But one stands out that I'll never forget. It wasn't what happened during the session that was memorable, but what happened afterward.

I went with the head of marketing for the company where I was consulting to interview a professional in our domain at her workplace. There she was, buried in a pile of folders, with calculators and calendars all around. I focused the discussion not on our product, but on understanding how she worked in general. By all accounts, it was a normal interview, or so I thought.

However, once we left the office building, the marketing person turned to me and said in a dry tone, "Our customers don't need our product." For all of the market research he'd done, he never had received that kind of firsthand insight. He hadn't considered what actually happened on the other end of his offering nor asked people how they thought about *their* process or needs.

From one single interview, my colleague was already thinking differently. And that was just a start. Imagine what we were able to uncover after a dozen more interviews. It turned out that our offering wasn't as important to our customers' needs as we thought. We weren't in their critical path. The company eventually learned that they had to find new ways to serve customers.

My experience reflects a key challenge: Had our head of marketing not witnessed the customer's problem firsthand, he wouldn't have had the same revelation he did about their needs. But not everyone in a company will get that chance. Indeed, most won't. So how, then, can we consistently translate insights about human needs into actionable intelligence?

Imprecise concepts like desires and emotions are hard to measure and quantify. Seeking to gain empathy, while well intended, lacks a clear beginning or end. It's no wonder that companies gravitate toward predictable and reliable research on market size and customer demographics. But traditional methods miss important, qualitative insight into why people act as they do.

JTBD provides a way to understand, classify, and organize otherwise irregular feedback. It not only directs you to look at your markets differently, but it also provides a clear and stable unit of analysis: the job.

JTBD lets you find the patterns that matter the most, taking the fuzziness out of the fuzzy front end of innovation.

Think of JTBD as an engine of inquiry that informs capabilities across departments—from innovation and strategy to product design and development to marketing and customer support. Having a common aim—understanding the core job and its related emotions and aspirations—is a necessary precursor to having aligned teams and efforts.

Defining JTBD

Every day, you have dozens of objectives that you strive to accomplish. You drink coffee to *get energy in the morning*. Then you might drive to a park-and-ride to take the train while you *commute to work*. At the office, you collaborate with colleagues to *complete a project* or *deliver a pitch* to a new client. Back home, you might eat a piece of chocolate to *reward yourself after work* and then *prepare a meal* to enjoy with your family.

These are all jobs to be done (JTBD).

The JTBD approach offers a unique lens for viewing the people you serve. Instead of looking at the demographic and psychographic factors of consumption, JTBD focuses on what people seek to achieve in a given circumstance. People don't "hire" products and services because of the demographic they belong to (e.g., 25–31-year-olds, have a college degree, earn a certain salary); instead, they employ solutions to get a job done.

JTBD is not about your product, service, or brand. Instead of focusing on your own solution, you must first understand what people want and why that's important to them. Accordingly, JTBD deliberately avoids mention of particular solutions in order to first comprehend the process that people go through to solve a problem. Only then can a company align its offerings to meet people's goals and needs.

Early origins of JTBD thinking point to Theodore Levitt. The famous business professor was known for telling students, "People don't want a quarter-inch drill, they want a quarter-inch hole."[1] This quote captures the essence of JTBD: focus on the outcome, not the technology. The drill is a means to an end, not the result.

Peter Drucker, a contemporary of Levitt and father of modern management, first used the phrase "jobs to be done" in relation to customer needs. In his 1985 book *Innovation and Entrepreneurship*, Drucker wrote:[2]

> Some innovations based on process need exploit incongruities, others demographics. Indeed, process need, unlike the other sources of innovation, does not start out with an event in the environment, whether internal or external. It starts out with the job to be done.

But neither Drucker nor Levitt used the label "job to be done" in any consequential way to refer to their ideas or approaches to solving business problems. It wasn't until Clayton Christensen popularized the term in *The Innovator's Solution*, the follow-up to his landmark work, *The Innovator's Dilemma*, that the concept became widespread.

Although modern references of JTBD point back to Christensen, definitions of JTBD vary in practice. Table 1.1 at the end of this chapter presents how a "job" is defined by thought leaders in the field. Comparing them side-by-side shows variation in approach, but also reveals commonalities.

Overall, JTBD is about understanding the *goals* that people want to accomplish, and achieving those goals amounts to *progress* in their lives. Jobs are also the motivators and drivers of behavior: they predict *why* people behave the way they do. This moves beyond mere correlation and strives to find causality.

My definition of a job is simple and broad:

> **The process of reaching objectives under given circumstances**

1. Although Levitt popularized this quote, he credits newspaperman Leo McGivena as his source of inspiration for the concept.
2. Peter Drucker, *Innovation and Entrepreneurship* (New York: Harper & Row, 1985), p. 69.

My use of the word "objectives" is deliberate. It better reflects the functional nature of JTBD. I don't use the word "goals" in my definition in order to avoid associations with broader aspirations, e.g., "life goals." This isn't to say that aspirations and emotions aren't important in JTBD. Instead, my interpretation of JTBD sequences the steps for creating offerings that people desire: first, meet the functional objectives and then layer the aspirational and emotional aspects onto the solution.

My definition of JTBD also includes an explicit mention of a process, highlighting the dynamic nature of getting a job done. In other words, an "objective" isn't just about an end point, rather an objective itself is a process that unfolds over time. The goal across the above definitions is the same: leverage a deeper understanding of how people make choices to create products they truly demand.

Perspectives of JTBD

Despite some common terminology and desired end results, the field of JTBD has unfortunately split into different schools of thought. Newcomers to JTBD may find an array of approaches and opinions on the topic, leading to confusion and discouragement. Contentious debates exacerbate the divide.

JTBD falls broadly into two camps. On the one side, there is the so-called "Switch" school of thought, pioneered by Bob Moesta. Through qualitative interviews, the Switch technique seeks to reverse engineer the underlying motivation for changing from one solution to another. The researcher can then deduce why people "hire" a solution to get a job done and analyze the forces of change. The aim is to increase demand for a given product or service.

On the other side is Tony Ulwick's Outcome-Driven Innovation (ODI), a strategy and innovation approach focused on pinpointing customer-centered opportunities. In qualitative interviews, ODI uncovers all of the desired outcomes that people want from getting a job done in a given domain. In a separate step, these desired outcomes are prioritized

MILKSHAKE MARKETING REVISITED

Harvard Business School professor Clayton Christensen often frames JTBD with a story involving milkshakes. He and his team were reportedly working with McDonald's to understand how to improve milkshake sales. Previously, the fast-food chain had tried changes to their milkshakes—making them thicker, chunkier, fruitier, etc. They also segmented consumers by demographics (e.g., age group) and tried to match those categories to different product variations. Milkshake sales did not improve.

Christensen's team took a different approach. Instead of focusing on product attributes, they went looking for the job to be done. Why do people "hire" milkshakes?

To start, they stood in a restaurant's parking lot and observed people getting milkshakes. Surprisingly, they learned that many people buy milkshakes in the morning on their way to work. The team dug deeper and asked people why.

It turned out that some people wanted to avoid getting hungry later in the morning. Having a snack on the drive to work helped fill them up, and milkshakes got that job done better than other options. Bananas were messy, bagels too hard to eat, a Snickers bar was unhealthy, and so on.

with a quantitative survey. ODI increases the adoption of innovation by creating products that address unmet needs.

I believe these two sides are not mutually exclusive, and there is a place for both. Sometimes, it makes sense to understand people's objectives and needs from the bottom-up (i.e., ODI), for instance, when developing a new product or when redefining your market. Other times, it's appropriate to start with a particular product in mind and understand why people "hired" that product to get a job done (i.e., Switch).

In the end, techniques from both interpretations can help your organization shift its mindset from inside-out to outside-in. There is a common focus on the underlying objectives that people have,

As a result of this story, the term "milkshake marketing" has come to refer to this type of reframing. It's a shift in traditional marketing approaches typically centered on demographic segmentation. Instead, Christensen advises, focus on why people "hire" products and services to get a job done.

Though widely cited, the milkshake example comes with several issues. First, the starting point is an existing solution—a milkshake—and thus frames the market a priori. Focusing instead on the job, the problem might be better cast as *get breakfast on the go.*

Second, milkshake sales, in fact, did *not* improve as a result of this study. Of course, there were many potential variables and reasons why the fast food chain didn't follow the advice of Christensen's team. But it was not a success story per se. (Personally, I don't know why McDonald's doesn't offer a low-sugar protein breakfast shake based on this insight).

You'll likely come across the milkshake story in your exploration of JTBD. Just be aware that, while popular and widespread, the example has limitations in applying JTBD. The focus of the story is on demand generation of an existing product (buying a milkshake) not on understanding the underlying objective (getting breakfast on the go).

independent of a solution. Ultimately, your goal is to make products people want, as well as make people want your products.

To be transparent, my approach is a loose interpretation of ODI, but it is not presented here as practiced by Ulwick's firm Strategyn. I've also practiced Switch techniques, and I follow the work of Bob Moesta closely. In working with both approaches, I have found a growing group of practitioners who benefit from a wide variety of JTBD techniques. This book strives to bring both perspectives together under one broad practice of JTBD.

Overall, JTBD is not a single method: it's a lens, a way of seeing. JTBD lets you step back from your business and understand the objectives of the people you serve. To innovate, don't ask customers about their preferences, but instead understand their underlying intent. Ultimately, JTBD seeks to reduce the inherent risk in innovation and ensure product-market fit from the outset.

There are many techniques that fall under the JTBD umbrella, and we'll look at many of the more popular ones that have surfaced over the past three decades throughout this book.

Principles of JTBD

At its core, JTBD is a way for organizations to look at needs and objectives rather than demographic and psychographic characteristics. JTBD theory predicts human behavior: individuals are motivated to make progress toward an objective. If an organization knows in advance what drives customer behavior, it has a better chance at creating successful solutions. Regardless of technique or interpretation of JTBD, there are five common principles many people in the field agree upon.

1. **People employ products and services to get *their* job done, not to interact with *your* organization.**

   JTBD doesn't look at the relationship that people have to a given solution or brand, but rather how a solution fits into their world. The aim is to understand their problems before coming up with solutions.

   To be clear, JTBD is not about customer journeys or experiences with product, which assume a relationship to a given provider. Customer journey investigations seek to answer questions such as: When do people first hear about a given solution? How did they decide to select the organization's offerings? What keeps them using it? These are all important questions to answer, but they also don't get to the underlying job.

   JTBD, on the other hand, focuses on the relationships that people have with reaching their own objectives. A given solution may or may not be employed in the process, but the job exists nonetheless, independent of any one provider. From this perspective, companies should also consider whether or not customers even want what they provide. Innovation often comes when a current means to an end is avoided altogether or absorbed into another process, thus eliminating the rationale for having the product or service to begin with.

2. **Jobs are stable over time, even as technology changes.**

   The jobs people are trying to get done are not only solution agnostic, but they also don't change with technology advancements. References to solutions (products, services, methods, etc.) are carefully avoided in JTBD vernacular. Consequently, JTBD research typically has a long shelf life. It is foundational insight that can be applied across projects and departments over time.

   For instance, 75 years ago when people prepared their taxes, they used pen and paper for all calculations and submissions. Later, they used pocket calculators to help with the numbers and sums. These days, completing taxes is done with sophisticated software and online filing solutions that didn't exist 50 years ago. Though technology changed, the job remains the same: *file taxes.*

3. **People seek services that enable them to get more of their job done quicker and easier.**

   New opportunities come from investigating the process of what people are trying to achieve. Mapping the job, not the buying journey, provides unique insight. Customers value getting a job done better.

   The Apple digital music ecosystem, for example, allows music enthusiasts to streamline how they *listen to music.* Not only can they listen to music on an iPod or iPhone, but they can also acquire and manage music with the iTunes system. Integrating various jobs—acquiring, managing, and listening to music—all in a single platform provided incredible market advantage. These days, streaming music services get that job done even better, but the job is the same.

4. **Making the job the unit of analysis makes innovation more predictable.**

   In a time when businesses are encouraged to "fail fast" and "break things," JTBD offers a more structured way to find solutions that resonate with customers in advance. Although there is no guarantee, understanding individuals' objectives and needs provides more targeted insight from the beginning. Product success isn't just left to luck or experimentation.

   Additionally, making factors like "empathy" the unit of analysis, as seen in Design Thinking, is problematic. When does empathy begin and end? How do you know when teams have achieved empathy? Instead, JTBD provides a concise focus: the job as an objective. Aspects like empathy, emotions, and personal characteristics of users can then be added later in a second phase when developing a particular solution.

5. **JTBD isn't limited to one discipline: it's a way of seeing that can be applied throughout an organization.**

   JTBD detaches up-front understanding from implementation. It gives a consistent, systematic approach to understanding what motivates people. As a result, JTBD has broad applicability inside of an organization, beyond design and development. Various teams inside an organization can leverage JTBD:

   - Sales can leverage JTBD thinking in customer discovery calls to uncover the objectives and needs that prospects are trying to accomplish.
   - Marketing specialists can create more effective campaigns around JTBD by shifting language from features to needs.
   - Customer success managers can use JTBD to understand why customers might cancel a subscription.
   - Support agents are able to provide better service by first understanding the customer's job to be done.

- Business development and strategy teams can use insight from JTBD to spot market opportunities, e.g., to help decide the next acquisition target.

Ultimately, JTBD can guide decisions and help craft solutions for any aspect of the business.

Benefits of JTBD

Overall, JTBD provides a human-centered way of viewing people you serve. The approach lets you connect with customers on their own terms. Use JTBD to focus your business on customer needs for improved performance and success.

JTBD is a foundational activity that enjoys longevity. Findings can be valid for years to come, helping you avoid the volatility of opinion-based research. As a result, you should find it is easier to future-proof your offering: since jobs are stable in time, they typically don't change at a rate faster than solutions do.

More importantly, JTBD shows causality: people act and decide in ways that help them achieve their objective. This, in turn, reveals real opportunities. Your ultimate aim is to use jobs thinking to find those solutions that have a good product-market fit and increase demand.

The effect is that JTBD helps break down silos between units within an organization. The common language fosters cross-departmental collaboration and aligns different teams to consistent targets. JTBD can be part of an overall mindset shift and cultural transformation.

Finally, JTBD is compatible with modern techniques, such as Design Thinking, Agile, and Lean. For instance, take an unmet need from JTBD research and turn it into a "how might we…" statement to kick off empathy-building exercises and ideation. Or user stories in Agile could be generated and organized based on jobs. Lean experiments could be framed around hypotheses statements that are grounded in JTBD research.

Recap

A job to be done is an objective that someone is trying to achieve in a given context. It's not about your product, solution, or brand, but what people want to accomplish. Thinking about how customers perceive value shifts your focus from inside-out to outside-in.

Precursors to JTBD go back to Theodore Levitt, who told his students, "People don't want a quarter-inch drill, they want a quarter-inch hole." Peter Drucker was the first to use the term "job to be done" in conjunction with what he called a "process need," or objective that people wanted to accomplish.

Clayton Christensen is universally credited with popularizing the concept of JTBD. But divergent schools of thought have divided the field into two camps. On one side is the Switch technique, which reverse-engineers motivations from a purchase experience. On the other is ODI, a comprehensive technique for determining business opportunity through unmet needs. Contentious debates between proponents of each side cause newcomers to experience confusion and distaste.

Regardless of the point of view, common core principles hold JTBD together as a field:

- People want to get a job done, not to interact with an organization.
- Jobs are stable over time.
- People seek services that help them get more of their job done, better.
- The job predicts behavior and becomes the key unit of analysis.
- JTBD isn't limited to one discipline; it applies across the organization.

The benefits of JTBD are many: JTBD shows causality, enjoys longevity, reflects a human-centered approach, helps break down silos and shift mindsets, and fits with modern techniques.

JTBD provides a consistent language for understanding people's motivations to reach an objective. The approach is not a single method or technique, but rather a way of seeing. This book gathers together many of the common approaches that have developed over the last 30 years of JTBD research and practice.

TABLE 1.1 COMPARING DEFINITIONS OF A JOB TO BE DONE

SOURCE	DEFINITION
Clayton Christensen, Taddy Hall, Karen Dillon, and David S. Duncan, *Competing Against Luck* (New York: HarperBusiness, 2016)	"A job to be done is your customers' struggle for progress and creating the right solution and attendant set of experiences to ensure you solve your customers' jobs well."
Bob Moesta, "Bob Moesta on Jobs-to-be-Done," interview by Des Traynor, *Inside Intercom* (podcast), May 12, 2016	"A job is really the process of making progress... It's helping them understand the struggles they have to go through to get to the progress they want... Remember, it's not Jobs—it's Jobs-to-Be-Done. It's about the thing they want to do better, and that's where innovation has to be."
Anthony Ulwick, "What Is Jobs-to-be-Done?" *JTBD+ODI* (blog), February 28, 2017	"The theory is based on the notion that people buy products and services to get a 'job' done. A 'job' is a statement of what the customer is trying to achieve or accomplish in a given situation."
Sandra M. Bates, *The Social Innovation Imperative* (New York: McGraw-Hill, 2012)	"Jobs are defined as the goals and objectives that people want to accomplish or what they are trying to prevent or avoid... Jobs are what motivate people to buy a product or service such as an iPhone, which enables them to 'be productive while on the go,' or auto insurance so they can protect against financial loss in the case of an accident."
Lance Bettencourt, *Service Innovation* (New York: McGraw-Hill, 2010)	"What the customer values is the ability to get a job done well. The customer job therefore offers a stable, long-term focal point for the improvement of current services or the creation of new-to-the-world services. Ultimately customers are loyal to the job."
Mike Boysen, "What #JobsToBeDone Is, and Is Not," *Medium* (blog), December 2017	"A Job is a goal or objective; or a problem that must be solved in order to create a desired future-state. Yes, it's progress as we are moving from a current-state to a future-state (in getting the job done). Executing a process or Job is progress. Solving problems is progress. Achieving goals and objectives is progress."
Stephen Wunker, Jessica Wattman, and David Farber, *Jobs to Be Done: A Roadmap for Customer-Centered Innovation* (New York: AMACON, 2016)	"While jobs are the tasks the customers are looking to get done in their lives, job drivers are the underlying contextual elements that make certain jobs more or less important."
Alan Klement, *When Coffee and Kale Compete,* 2nd Ed. (Self-published, 2018)	"A Job to Be Done is the process a consumer goes through whenever she evolves herself through buying and using a product. It begins when the customer becomes aware of the possibility to evolve. It continues as long as the desired progress is sought. It ends when the consumer realizes new capabilities and behaves differently, or abandons the idea of evolving."
Des Traynor in *Intercom on Jobs-to-be-Done,* (Self-published, 2016)	"Jobs-to-be-Done... lets you focus on making things people actually want. When you're solving needs that already exist, you don't need to convince people they need your product."

CHAPTER

2

Core Concepts of JTBD

IN THIS CHAPTER, YOU WILL LEARN:

- How to separate the different elements of JTBD
- Guidelines for formulating JTBD
- The hierarchical nature of JTBD
- How to get started scoping a JTBD project

In 1543, Nicolaus Copernicus published a startling finding about our solar system. In *De revolutionibus orbium coelestium* (*On the Revolutions of the Celestial Spheres*), he proved mathematically that the earth revolves around the sun and not the other way around, as previously believed. This represented a paradigm shift away from thinking that the earth was the center of the universe.

To explain his heliocentric theory, Copernicus created a model of the solar system, shown in Figure 2.1. Of course, the movement of the planets is more complicated than shown. The orbits of the planets are not perfect circles, for one, and the distances between them aren't nearly as uniform as suggested in his visualization. His model is an inaccurate abstraction but makes a clear point: the earth is not in the center of our solar system, the sun is.

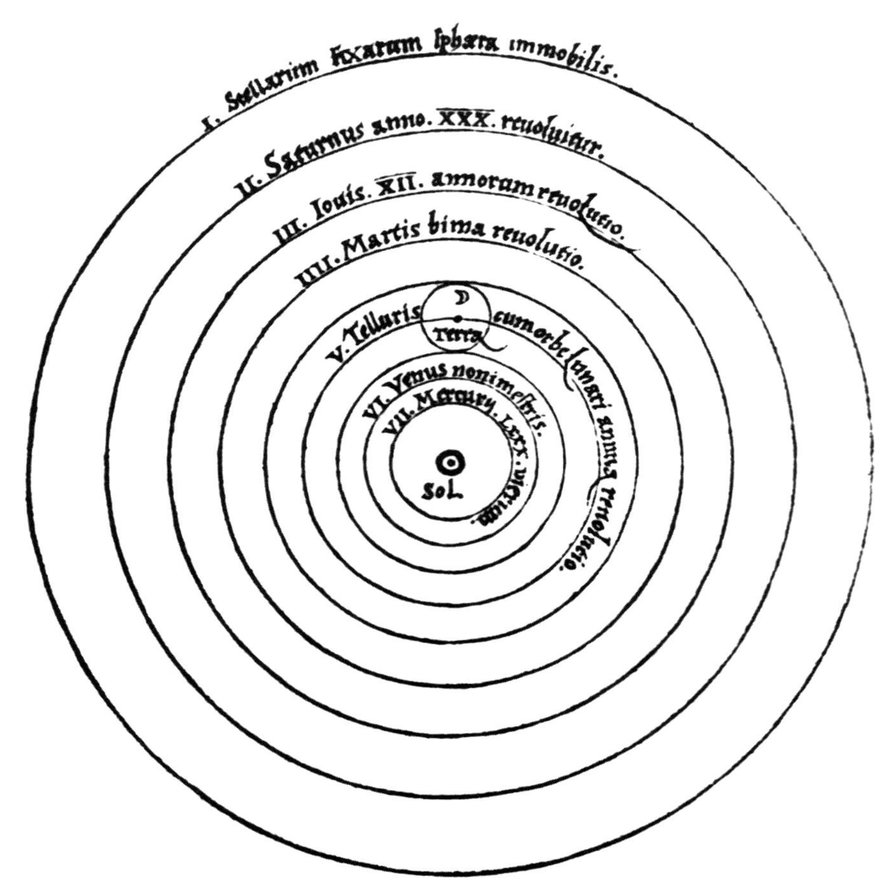

FIGURE 2.1 Copernicus reflected his observations in a model of the solar system, putting the sun at the center instead of the earth.

Although the ensuing Copernican Revolution took decades to take hold, his discovery wasn't just of scientific importance. It also shook the mindset of people at that time: no longer could people see themselves as the center of the universe.

In some sense, we're witnessing a type of Copernican Revolution in business. Instead of organizations and their brand products being in the middle, the customer stands at the center of the business universe now. After all, the competition is just one click away, and customers have an increasing number of options. The business imperative has flipped from a *push* to a *pull*: companies don't sell products, they buy customers. To do so, companies have to understand the fundamental needs and objectives of people in their market.

The problem is that many businesses aren't ready to absorb the effects of this paradigm shift. They instead cling to management models of the past, despite the new customer-centric imperative. They struggle to find a new center to their view of the world. But markets don't wait: value isn't measured by some feature set or capability, but rather how people perceive the benefits of an offering.

JTBD helps. It provides a systematic framework for creating your own model of people's needs. Like Copernicus's diagram, models of the job to be done are abstractions. But those abstractions are an important foundation for integrating human needs into business decision-making.

To understand how JTBD can help shift mindsets, let's first look at the various elements of the framework and how they come together to provide a new North Star to follow.

Elements of JTBD

A core strength of JTBD is its structure, which clearly separates out various aspects of achieving objectives. The who, what, how, why and when/where are analyzed individually, giving both precision and flexibility to JTBD methods. My simplified model of JTBD consists of five core elements, illustrated in Figure 2.2.[1]

- **Job performer** (who): The executor of the main job, the ultimate end user
- **Jobs** (what): The aim of the performer, what they want to accomplish
- **Process** (how): The procedure of how the job will get done
- **Needs** (why): Why the performer acts in a certain way while executing the job, or their requirements or intended outcomes during the job process
- **Circumstances** (when/where): The contextual factors that frame job execution

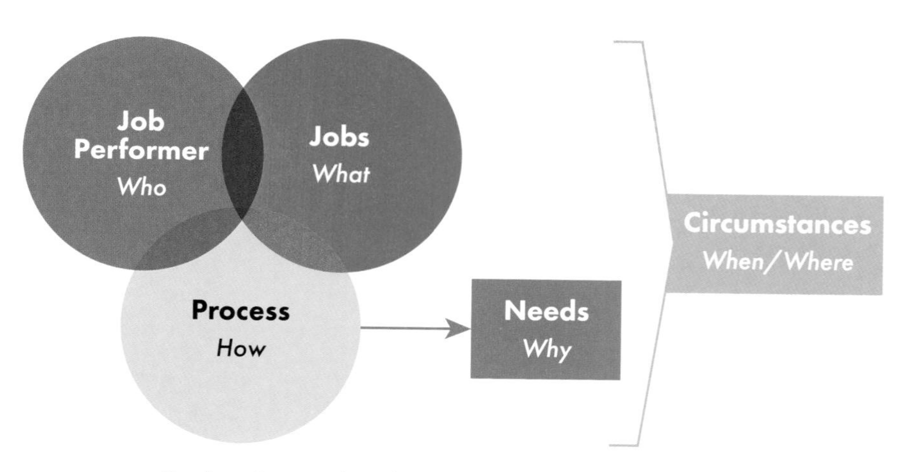

FIGURE 2.2 Five key elements describe the JTBD ecosystem.

1. Although my model resembles Tony Ulwick's ODI framework, it is an interpretation of the approach and not the same as practiced by his company Strategyn.

Job Performer

Who is trying to get the job done? The *job performer* represents the individual who will be executing the job. That person is the eventual end user of the services you'll provide.

Be sure to make a distinction between the various functions involved in performing the job, in particular differentiating the *performer* from the *buyer*. Don't conflate the two, because they have different needs. Think about two separate hats that are worn: one is for the job performer while carrying out the job; the other is for the buyer when purchasing a product or service.

Now, in B2C contexts, a single person may switch between the two hats. But their needs while wearing each hat are distinct. In the B2B situations, the job performer and the buyer are often separate people. For instance, a procurement office may purchase equipment and materials for others in the company without their direct input.

In addition to the *job performer* and the *buyer*, other functions within the job ecosystem to consider include the following:

- **Approver:** Someone who authorizes the acquisition of a solution, e.g., a controller, a spouse or parent, or a budget holder
- **Reviewer:** Someone who examines a solution for appropriateness, e.g., a lawyer, a consultant, or a compliance officer
- **Technician:** The person who integrates a solution and gets it working, e.g., an IT support, an installer, or a tech-savvy friend
- **Manager:** Someone who oversees a job performer while performing the job, e.g., a supervisor, a team lead, or a boss
- **Audience:** People who consume the output of performing the job, e.g., a client, a downstream decision-maker, or a team
- **Assistant:** A person who aids and supports the job performer in getting the job done, e.g., a helper, a teammate, or a friend

Map out the different actors who may be involved in a simple diagram, such as the one shown in Figure 2.3.

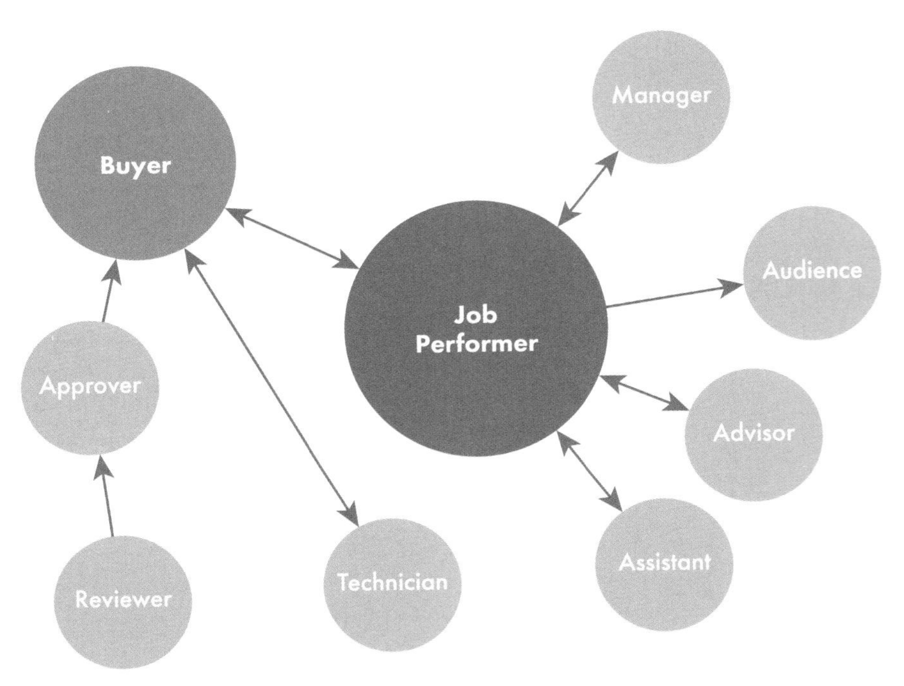

FIGURE 2.3 Keep the possible functions in a job ecosystem separate and focus on the job performer first.

Note that these different roles don't refer to job titles. Instead, they represent different functional actors within the context of getting a job done. To reiterate, consider these as separate roles or hats to wear. The primary focus on attention is on the job performer first. Later, you can consider the needs of the other roles in relation to the job to be done and the job performer.

For instance, let's say your company provides an online task management tool to enterprises. The *job performer* is the ultimate end user, perhaps a programmer on a development team. The *buyer* could be the collaboration software manager of the customer's company, for example, who may need to get *approval* from a procurement office and have the legal department *review* any software agreement. The job performer also has a *manager*, who determines the practices around assigning tasks. Program managers may be the *audience* of the job performer when she presents progress to them.

Draw a quick map, like the one in Figure 2.3, to help differentiate the job performer from other roles. The JTBD perspective gives a sequence in which to address the needs of various stakeholders: start with the needs of the job performer; then focus on the buyer before looking at the needs of others. Keep in mind this doesn't mean that buyer needs are unimportant. Instead, solutions must first and foremost address the needs of the job performer ahead of considering the needs of the purchaser.

Jobs

What is the job performer trying to achieve? A job is a goal or an objective independent of your solution. The aim of the job performer is not to interact with your company but to get something done. Your service is a means to an end, and you must first understand that end.

Because they don't mention solutions or technology, jobs should be as timeless and unchanging as possible. Ask yourself, "How would people have gotten the job done 50 years ago?" Strive to frame jobs in a way that makes them stable, even as technology changes.

There are several types of jobs you'll ultimately be looking for, and it takes practice to sort them out and define them. The key distinctions to make are between the main job, related jobs, and emotional and social jobs.

MAIN JOB

The main job is the overall aim of the job performer. Determining the main job defines your overall playing field and sets your scope of innovation. You should express the main job in functional terms, such as a utilitarian goal. It's an act that will be performed and should have a clear end state—the "done" part of jobs to be done.

The main job is broad and straightforward, serving as an anchor for all other elements of your JTBD investigation. For example, *prepare a meal*, *listen to music*, or *plan long-term financial well-being* are examples.

The main job shouldn't include adjectives like *quick*, *easy*, or *inexpensive*. Those are considered to be needs, or the metrics by which job performers compare solutions, which are handled separately. The main job is also different from your marketing message or value proposition statement, which tends to be persuasive to evoke an emotion.

Figure 2.4 illustrates the relationship of the different types of jobs, with the main job framing the primary scope of inquiry. It may be necessary to look at related jobs when devising a solution, or to consider broader jobs or narrower jobs. The main job sets your focus and everything else is seen in relation to it.

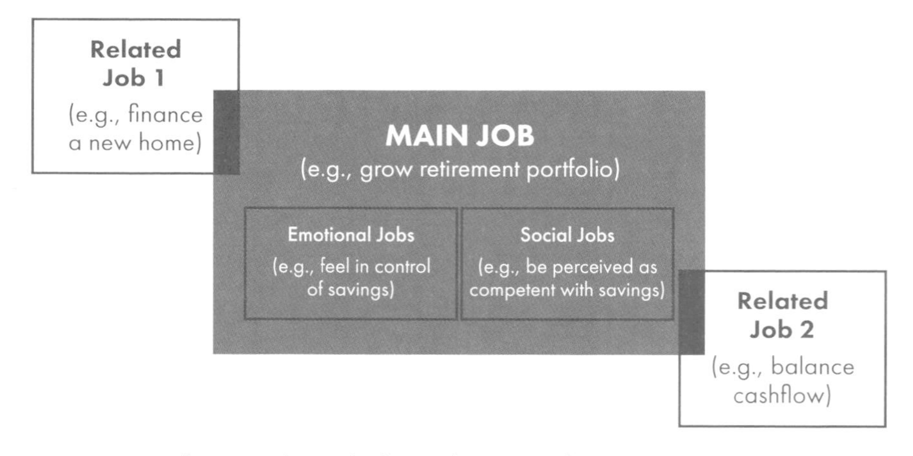

FIGURE 2.4 The main job sets the focus of inquiry and innovation with other aspects in relation to it.

RELATED JOBS

Related jobs are adjacent to the main job, but are significantly different. For instance, if you define *grow retirement portfolio* as a main job, related jobs may be *finance a new home* or *balance cash flow*. Identifying related jobs as such can help your team understand the main job—what it is and what it is not.

Also recognize that people have multiple goals that collide and intersect. The world is not as neat and compact as your JTBD model will suggest. As you define the main job, identify related jobs to understand the overall landscape of objectives. Only then should you decide on a single main job to focus on, keeping related jobs in your peripheral line of sight.

Keep in mind that related goals may even compete with the main job and each other. For instance, buying a large-ticket item like a car or house may detract from growing a retirement portfolio. As a result, progress in our lives is the sum of the outcomes of related jobs, and balance is often required.

EMOTIONAL AND SOCIAL JOBS

Emotional jobs reflect how people want to feel while performing the job. Statements usually start with the word "feel." For example, if the main job of a keyless lock system is to *secure entryways to home,* emotional jobs might be to *feel safe at home* or *feel confident that intruders won't break in while away.*

Social jobs indicate how a job performer is perceived by others while carrying out the job. For instance, adult diapers have an important social job of *avoiding embarrassment in public*. Or, in the previous example, the person with a keyless door lock might be seen as an innovator in the neighborhood.

Separating functional jobs from emotional and social jobs helps focus on the individual's objective, on one side, and experiential aspects of getting the job done, on the other. The rule of thumb is to solve for the functional job first. It's hard to solve for an emotional or social job if the functional job isn't fulfilled.

I saw this firsthand on a project I once led in the area of online women's fashion. Most of the discussion revolved around emotional and social jobs of fashion, e.g., *feel confident in public* or *look good to others.*

But we found that the key unmet need when purchasing clothes was whether the item fit—a functional job. Even in a domain where emotions dominate, our focus turned to solving the functional job first, not the emotional or social job.

The point is that JTBD provides a sequence for innovation: meet the needs of the functional job first and then layer emotional and social aspects after that. Targeting emotional and social jobs first often yields an endless number of solution possibilities. There are many ways to help customers feel confident in public, for instance. Starting with a functional job grounds innovation in concrete options that are feasible, but emotional and social jobs are not overlooked.

FORMULATING JOB STATEMENTS

The value of JTBD lies in its consistent way of describing people's goals and needs. Keep in mind that a job is not what your organization needs to do to deliver a service: it's what the job performer wants to get done. Your team's tasks are not your customers' goals. Always think in terms of the individual's perspective.

To be consistent in describing goals, follow this simple pattern for writing job statements.

> *verb + object + clarifier*

Examples include: *visit family on special occasions, remove snow from pathways, listen to music on a run*, and *plan long-term financial well-being.* Keep in mind that needs are handled separately, so typically adjectives are omitted that qualify how well jobs get done.

Formulating job statements takes practice. To provide a common language for your organization, getting the words and syntax right is important. One trick is to think of a silent "I want to…" in front of each statement that then gets omitted later. Also don't include other phrases like "help me…" at the front of the job statement. Instead, begin directly with a verb.

For clarity, it's also possible to include examples after a job statement. Use the abbreviation "e.g." to add some specific instances that are representative of the types of jobs you're describing. For instance, you could qualify a job *visit family on special occasions* with *e.g., a birthday, graduation, marriage, or holiday.*

Table 2.1 outlines the guidelines for formulating job statements.

TABLE 2.1 RULES FOR FORMULATING JOB STATEMENTS

DOS	DON'TS
Reflect the individual's perspective	Never refer to technology or solutions
Start with a verb	Steer clear of methods or techniques
Ensure stability over time	Don't reflect observations or preferences
Clarify with context, if needed	Avoid compound concepts (no ANDs or ORs)

Table 2.2 shows some examples of incorrect job statements and the issue with their formulation. A better expression is provided in the column on the right following the above guidelines.

TABLE 2.2 EXAMPLES OF FORMULATING JOB STATEMENTS

INCORRECT	ISSUE	CORRECT
Search by keyword for documents in the database	Includes specific methods (keyword search) and technology (documents in database)	Retrieve content
People prefer to attend meet-ups and conferences that are nearby	Reflects an observation and preference Includes compound concepts	Attend an event
Find the cheapest airfares quickly	Includes needs, which should be considered separately (i.e., cheap and quick)	Find airfares
Help me plan a vacation that the whole family will enjoy	"Enjoy" is a need and should be separate Includes "help me" instead of starting with a verb	Plan family vacation

Jobs exist independent of your solution or offering. They are discovered through qualitative research, discussed in the next chapter.

Process

How does the job get done? JTBD sees an "objective" as a procedure or a process. Job performers move through different stages of the goal as they strive to accomplish it. Understanding the process of the job performer's intent is key to JTBD.

You can illustrate the main job in a chronological map with a sequence of *stages*. Consider each stage as a smaller job within the main job rather than tasks or physical activities. Because the job has to be "done," be sure to formulate the job in a way that has an end state. It's then helpful to think of the job as having a beginning, middle, and end stage as well.

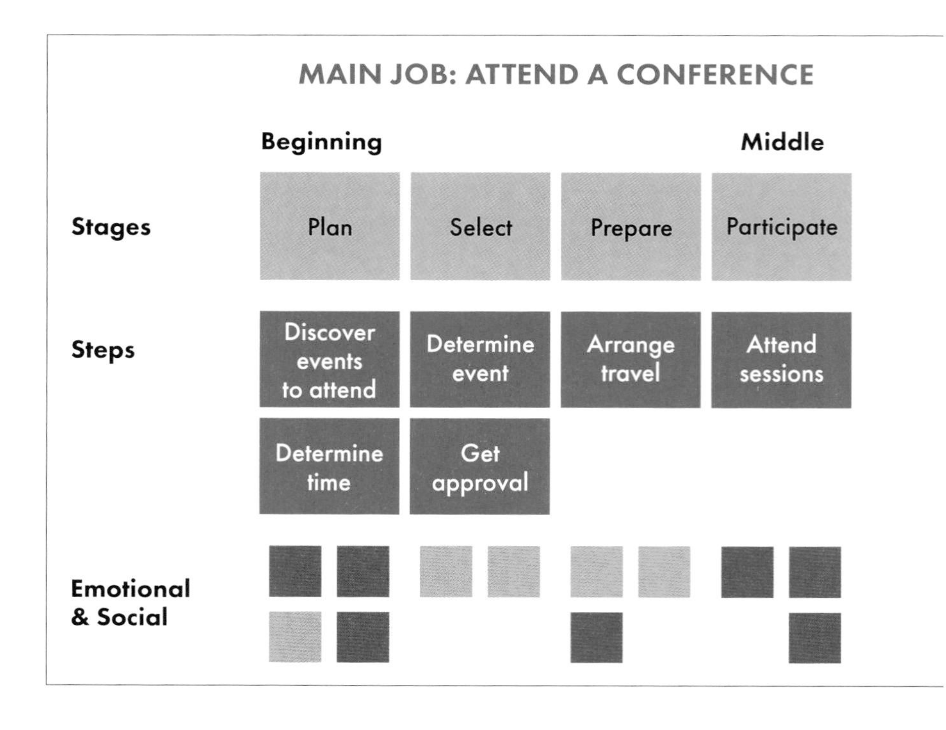

Once you have the main sequence, specify smaller *steps*. Note that steps are not tasks, but the smallest discrete subjob in the process. The guidelines for job formulation in Table 2.1 apply to job steps, too.

It's also possible to locate emotional and social aspects in a map of the main job. In the end, the map serves as a centerpiece in the JTBD framework that allows you to organize your description of the job. Later, you can use the job to organize needs as well. It becomes a central structure for compiling insight and focusing team conversations.

Using the main job, *attend a conference* as an example, the basic sequence could be visualized as shown in Figure 2.5.

It's critical to recognize that a job map is not a customer journey map. The aim is not to document how people come to your solution, decide to purchase, and stay loyal. That's not *their* job to be done; it's what your company wants them to do. Instead, a job map is a view into the

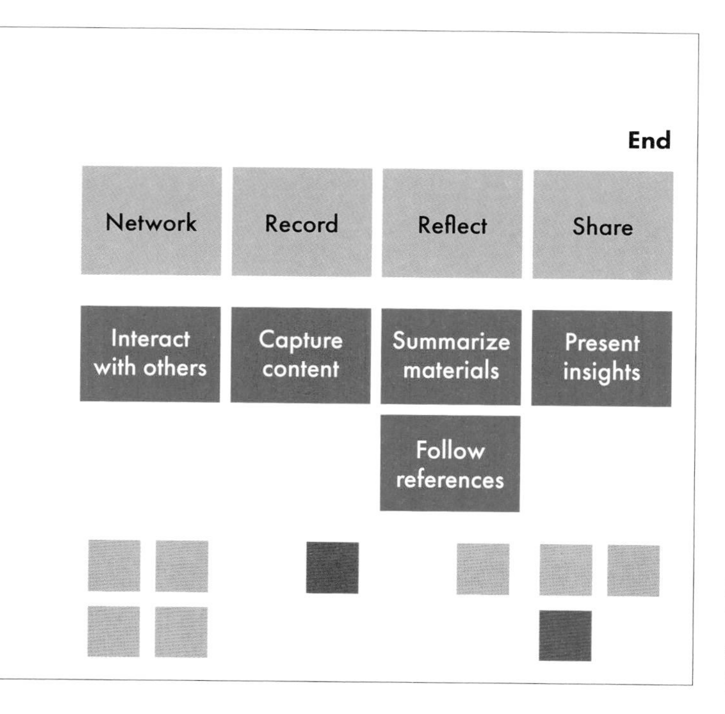

FIGURE 2.5
Visualize the process of getting a main job done.

behaviors and needs of individuals in the context of their daily lives. That may or may not include your solution.

A job map not only lets you see the bigger picture for strategic opportunities, but it also gives you a way to examine specific points that give rise to specific innovative ideas to fulfill a customer's job. In some cases, the map provides enough insight on its own to get started finding the right solutions. For instance, a startup looking to experiment with solutions can use the job map to align its features and functions to the job.

The process of creating a job map is described in detail in Chapter 3, "Discovering Value."

Needs

Why do the job performers act the way they do while getting the job done?

Working with needs is challenging, in general. The term itself has many connotations and defies precise definition. In some cases, such as in software development, a "need" is interpreted as a system requirement. For instance, user stories in Agile methodologies describe what users need to have in place in order to interact with a system.

In other cases, such as with voice of the customer research, a "need" is a benefit a customer gets from a given product or solution, e.g., customers need to have quick and easy access to support. And in yet other situations, such as with Design Thinking techniques, a need is seen as a fundamental human motivation, e.g., people need personal fulfillment. Without a common understanding and clear definition of what a need is, conversations within teams and organizations can go astray quickly. It's no wonder there isn't agreement on what a need is exactly.

JTBD helps in two ways. First, in JTBD a "need" is seen in relation to getting the main job done. Needs aren't demands from a solution, but an individual's requirements for getting a job done. For instance, if a main job is defined as *file taxes*, needs in getting that job done may be *minimize the time it takes to gather documents* or *maximize the likelihood of a getting a return*.

From this standpoint, needs aren't aspirations either, which are above the main job in terms of abstraction. Expressions like "have financial peace of mind" or "provide for my family" are motivations beyond getting the main job. These are important aspects to consider later, but not needs related to reaching the objective of filing taxes.

Second, JTBD provides a consistent pattern for expressing needs, as described earlier in the chapter: *verb + objective + clarifier*. This normalization allows for consistent ways of pinpointing opportunity. Semantics matter. Without a clear, concise formulation of what a need is, interpretation is up in the air.

Think of the job as the overall object or aim and needs as the success criteria along the way. As with job statements, formulating a need statement in a standard way is critical. Lance Bettencourt and Anthony Ulwick have developed a consistent way to notate needs in what they call *desired outcome statements*. There are four elements:

Direction of change + unit of measure + object + clarifier

- **Direction of change:** How does the job performer want to improve conditions? Each need statement starts with a verb showing the desired change of improvement. Words like "minimize," "decrease," or "lower" show a reduction of unit of measure, while words like "maximize," "increase," and "raise" show an upward change.
- **Unit of measure:** What is the metric for success? The next element in the statement shows the unit of measure the individual wants to increase or decrease. Time, effort, skill, and likelihood are a few typical examples. Note that the measure may be subjective and relative, but it should be as concrete as possible.
- **Object of the need:** What is the need about? Indicate the object of control that will be affected by doing a job.
- **Clarifier:** What else is necessary to understand the need? Include contextual clues to clarify and provide descriptions of the circumstance in which the job takes places.

Consider the following examples of need statements in Table 2.3 for the job *attend a conference*.

TABLE 2.3 EXAMPLE NEED STATEMENT FORMULATIONS

DIRECTION	MEASURE	OBJECT	CLARIFIER
Increase	the likelihood	of getting permission	from a boss to attend
Maximize	the ability	to remember relevant content	from conference presentations
Minimize	the time it takes	to summarize conference insights	for sharing with colleagues
Maximize	the likelihood	of networking with thought leaders	in the field

A strength of the JTBD approach is that it separates goals from needs. Consider the job as the target that someone wants to get done, and the need as the measure of success or the expected outcome. For instance, the statement "speed up the next big promotion at work" mixes goals and needs. In JTBD terms, the job is simply to *get a promotion*. Needs in completing that job are to *minimize the time it takes to get promoted* and *increase the size of the promotion*.

If you find yourself using an *and* or *or*, you probably need two separate need statements. Need statements should be as atomic as possible to not only see all of the factors involved in getting a job done, but also to pinpoint which ones are most important. As a result, any main job may have 50–150 intended needs.

For instance, there may be many needs for the main job of *prepare a meal*, such as *minimize time to process ingredients*, *reduce the risk of injury*, *increase the likelihood that others will enjoy the meal*, or *minimize the effort to clean up after the meal*. The list goes on.

Circumstances

When and where does the job get done? JTBD also takes the context of getting the job done into account in order to be relevant to an organization. For instance, *get breakfast* is a very broad job that could apply to many situations. But for a fast food restaurant, *get breakfast on the go*, is a more precise job to focus on. The conditions around the job give it meaning and relevance and therefore must be considered.

Adding contextual detail to the situation also helps greatly when designing a solution. For instance, a solution for the job *get breakfast on the go* could include everything from going to a restaurant or diner to eating a packed lunch at a desk. But when considering specific circumstances like *when late for work*, *while commuting* and *when cost is a factor*, a morning milkshake might be a better solution for the job.

Circumstances typically consist of aspects around time, manner, and place. For example, for the job *listen to music on a run*, you might uncover factors that determine how that will be performed:

- When it's raining
- When there is traffic on the road
- When it gets sweaty
- When it's bouncy
- And so on…

A job without context is not complete and cannot provide strategic direction. There is necessarily a dependency on formulating a main job and knowing the circumstances.

To summarize, there are five key elements in the JTBD framework. Using the previous example of attending a conference, you could consolidate an expression of the JTBD elements as follows in Table 2.4.

TABLE 2.4 AN EXAMPLE OF JTBD ELEMENTS

ELEMENT	EXAMPLES
Job Performer	Conference attendee
Main Job	Attend a conference, e.g., a conference, symposium, meetup
Related Job	Earn continuing education credits Take a training course
Emotional and Social Jobs	Feel inspired by new information Network with others who are like-minded
Process	Main stages: 1. Plan 2. Decide 3. Prepare 4. Participate 5. Network 6. Record 7. Summarize 8. Share
Needs	• Reduce the time it takes to select a conference to attend • Maximize chances of getting permission to attend • Increase connections in a professional network • Raise awareness of the latest topics in the field • Minimize the time it takes to share learnings with others • Maximize the ability to recall relevant content from event • Etc. (up to ~150 such statements)
Circumstances	• When companies limit the number of events employees can attend • When it's the first time attending a conference • When it's a very small conference • When it's a very large conference • Etc.

Hierarchy in JTBD

The parable goes like this: A traveler came upon three stoneworkers arranging bricks and asked them what they were doing. The first replied, "I'm laying bricks." The second answered, "I'm building a room." When the traveler got to the third man, he heard a different response. "I'm building a cathedral," the stoneworker replied. Of course, all three answered correctly; it's just a matter of perspective.

When working with JTBD, you'll confront the issue of granularity. The question you need to answer is, "At what level of abstraction do you want to try to innovate?" There's no right or wrong answer—it depends on your situation and aim. Getting the right altitude is key. Objectives at one level roll up into higher-order goals, generally called *laddering*. In JTBD work, the principle of laddering applies as well. For instance, in his book *Competing Against Luck*, Clayton Christensen points to "big jobs," or things that have a big impact on our lives (like finding a new career) and "little jobs," or things that arise in our daily lives (such as passing time while waiting in line).

Rather than just two levels—big and little—I have found that JTBD can be viewed on about four different levels (see Figure 2.6).

- **Aspirations:** An ideal change of state, something the individual desires to become
- **Big Job:** A broader objective, typically at the level of a main job
- **Little Job:** A smaller, more practical job that corresponds roughly to stages in a big job
- **Micro-Job:** Activities that resemble tasks but are stated in terms of JTBD

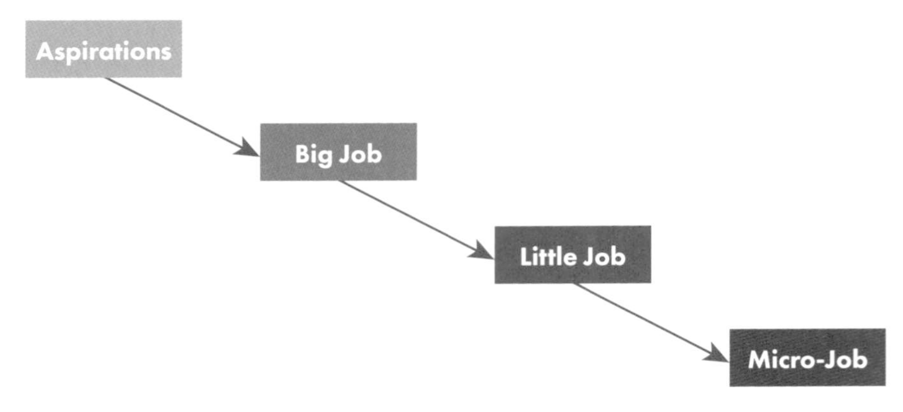

FIGURE 2.6 Recognize the different levels of abstraction in JTBD.

For example, I once experienced a mismatch in goal levels while interacting with Zipcar, the popular carsharing service. On one occasion, I showed up to my reserved car spot, and the vehicle wasn't there. After immediately calling customer service, they offered to pay for a cab ride to my destination. Job done.

But on Mother's Day another year, the same thing happened: I arrived to my reserved car, and it wasn't there. This time, rather than offering an alternative ride option, the Zipcar agent proceeded to name other locations to pick up a Zipcar. All of them would have added hours to my trip.

In other words, while I was trying to get to my family's house at a specific time (a big job), in this case, the agent was trying to rent me another car (a little job). Consider some of the different levels of abstraction in this example, reflected in Table 2.5.

TABLE 2.5 LEVELS OF JOBS TO CONSIDER

LEVEL	EXAMPLE
Aspiration	Be a better family member
Big Job	Visit with family on special occasions
Little Job	Arrange transportation at a specific time
Micro-Job	Start vehicle

There are many stages of the main job, *visit with your family*, including planning, scheduling, arranging, traveling, arriving, visiting, and departing, for instance. Each of these smaller jobs can be broken down further. Arranging transportation has the steps of deciding, reserving, confirming, and initiating, for example. Typically, the main job is broad to be more inclusive, but it can also be broken down into subsequent parts.

Note that the aspirations are technically *not* jobs. There are many ways to be a better family member, and there is no real "done" state. But oftentimes, your team may ladder up in defining the main job, gravitating toward overarching motivations. Having the category "aspirations" lets you capture a high-level thought, but then move down to the appropriate level of discussion.

For instance, if you find yourself defining the main job as *be a better professional* or *enjoy the arts* or *even be satisfied in life*, you probably need to move down a level in the hierarchy. Make a note of a relevant aspiration—it will help frame how a solution gets designed and marketed. But scoping your initial innovation effort at the aspirational level can yield an endless number of possible directions. It's more effective to target a big job and layer aspirations secondarily on top of that.

To reiterate, JTBD provides a sequence for innovation: start with the job performer and the main job defined at an appropriate functional level. Create solutions that get that job done first. Then consider aspects like emotions and aspirations for framing how the solution gets implemented and delivered to a market.

Level Set with "Why?" and "How?"

Keeping your work at the appropriate granularity can be tricky, but part of the territory. Sometimes, you need to know the broadest possible jobs—how customers want to change their lives. Other times, you'll be operating at a lower level with a narrower scope.

Two simple questions can help you get the right altitude: asking "why?" moves you up in the hierarchy; asking "how?" moves you down. (See Figure 2.7.)

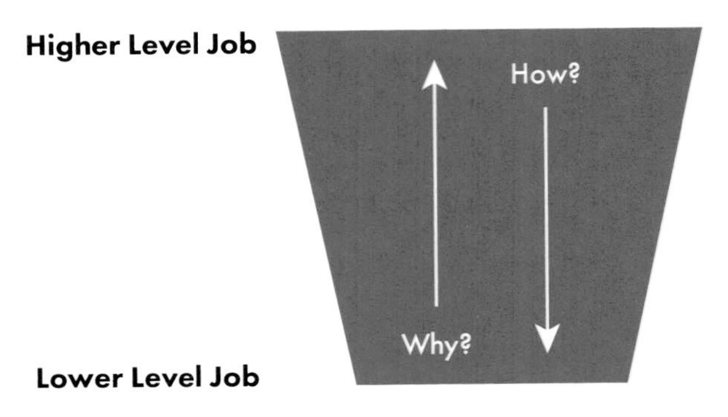

FIGURE 2.7 Ask "why?" to go up a level and "how?" to go down.

For instance, in the case of *attending a conference*, you might ask, "Why would a job performer want to go to a conference?" The answer might reveal that it's part of a broader aspiration around professional development. If you then ask again, "Why does the job performer want to develop professional skills?" you might find it's for career advancement and ultimately to have a better life. Those higher-level aspirations are good to know, and they have potential market appeal. But keep the functional job in mind first. If it doesn't get done, the aspiration won't be reached either.

On the other hand, asking, "How does the job performer attend a conference?" you might find the smaller job of convincing a boss to give permission. If you then ask, "How does the job performer convince the boss?" you might find a micro-job of providing a cost-benefit estimation of the event.

You can also use this technique—asking "why?" and "how?"—when interviewing people about their jobs to be done. See more in the next chapter about interviewing. Note that JTBD is not a game of asking the "5 Whys," a popular technique for root cause analysis that poses the questions "why?" successively five times. Instead, the level of innovation should match your goal, but be broad enough to allow for expansion.

Putting It All Together: Scope the JTBD Domain

Before getting started, you first have to define the target domain and the breadth of your field of inquiry. A main job can be big or small, and the focus depends on your situation. It's up to you to set the level of altitude and the boundaries of the job, discussed below.

Make scoping your JTBD domain a team activity. It's your first chance to get others in your organization on board to become more job-centric. Note that some people find formulating JTBD somewhat artificial and confining. The approach demands rigor that takes practice and some getting used to. Start small and practice as you go.

There are three key steps in scoping the JTBD domain: defining the main job, defining the job performer, and making a hypothesis about the process and circumstances. Each of these is guided by the business an organization is in currently, as well as the position it wants to occupy in the future. Your initial definitions may change as you learn more through in-depth research, but to begin it's best to be as targeted as possible initially.

Also, remember that to define each element of your JTBD model, you should speak with potential job performers. Avoid making assumptions, and ground your definitions in reality from the start. With just a few one-on-one interviews, you'll be able to learn a lot about the job performers and their job to be done.

Define the Main Job

What is your customer's primary objective you want to understand? You may not be able to answer this question right away. It takes negotiation and iteration. Work directly with your whole team to define a main job that makes the most sense. Discuss and refine the scope, get the right level and specificity, and then ensure the correct formulation of your main job statement.

The main job lies at the center of understanding a market. It becomes a centripetal force for making decisions and aligning to customer needs and desires. Related jobs adjacent to the main job point to further opportunities to serve customers. Drilling down into the steps of getting a job done provides insight into how to develop better products and solutions. Reaching upward to broader jobs and aspirations allows organizations to expand their businesses in general.

Getting the right level of abstraction is key. Don't define the main job too narrowly. A small job will limit your field of vision, but also will constrain your efforts. When in doubt, go broader and define a main job that is larger than smaller. Ask "why?" and "how?" to move the level of granularity of the main job up or down.

Consider how much time and effort you want to put into perfecting any one job relative to your size. JTBD thought leader Mike Boysen drives this point home when he picks on parking apps.[2] *Parking a vehicle* is such a small part of a much bigger job of *getting to a destination on time*, as shown in the job map in Figure 2.8.

Creating a parking app may be exactly the right place to start for a small firm or team. Taking on too much at once can be a recipe for failure. But then what? Focusing on getting just one small job done won't likely lead to long-term sustainability of a company. Instead, your strategy can expand by getting more steps done for customers. Defining the main job one level broader than your current capabilities provides an exit strategy and a path toward growth.

2. Mike Boysen, "If You Can't Identify an Exit Strategy, You Can't Identify Your Market," *Medium* (blog), May 6, 2017, https://mikeboysen.com/if-you-cant-identify-an-exit-strategy-you-can-t-identify-your-market-jtbd-d39961539618

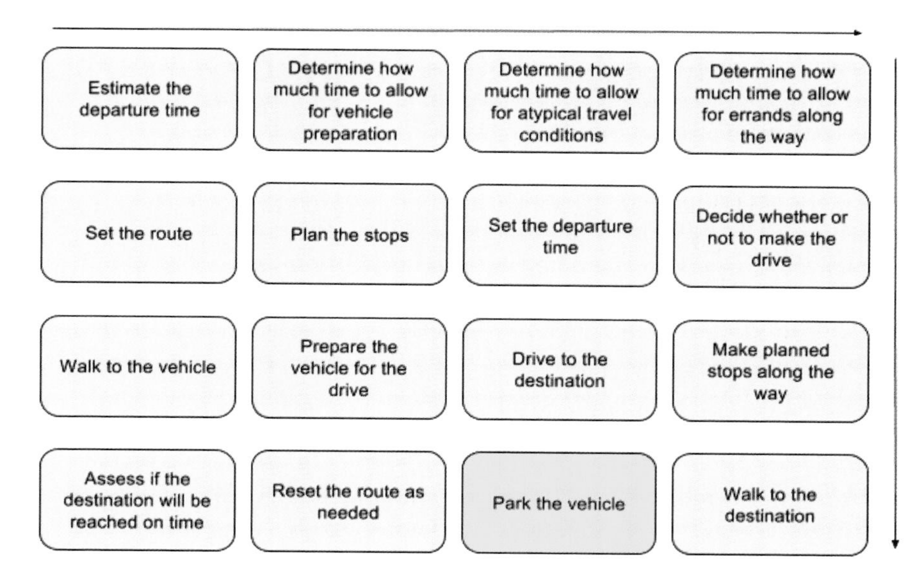

FIGURE 2.8 Illustrated in this job map, Mike Boysen shows that *park the vehicle* is a small step in a larger job *get to a destination on time.*

At the same time, avoid defining the main job as an aspiration or a description of an experience. "Be thoughtful" or "Be the best I can be" is too abstract: your innovation teams won't be able to act on it. It's okay to recognize aspirations, but keep the main job simple and functional. A rule of thumb is to focus on the "what" main job—a functional objective—more than the "why" in terms of aspiration.

Also strive to keep a causal impact of your business on the job. For instance, a drill bit manufacturer might explore a main job and realize that people don't want a hole, they want to hang a painting. But why do they do that? To decorate their homes and preserve family memories. Realistically, a team responsible for the development and sale of drill bits won't have a direct impact on creating a better home life, and practical innovation won't happen with a focus on a main job that is causally separated from the business.

Reflect on a few simple questions to get started:

- What business are you in? Consider your overall playing field by noting the sector, industry, and offering category you want to operate in.
- What customer problems do you want to solve? Write down all of the challenges you want to overcome for customers.
- What impact do you hope to generate? Write down the benefits you hope to bring customers.

Then create a simple ladder of objectives that people are trying to achieve, similar to the one in Table 2.6. Using the example of parking a car, you can see how this job might fit into a hierarchy of jobs.

TABLE 2.6 GET THE RIGHT LEVEL OF ABSTRACTION FOR THE MAIN JOB

LEVEL	EXAMPLE
Aspiration	Enjoy the freedom of mobility
Big job	Get to a destination on time
Little job	Park the vehicle
Micro-job	Find a free parking spot; feed the meter, etc.

If you already have access to people in your domain, talk with them to explore the main job informally. A short phone call or quick meeting with a few known individuals carrying out the job will help you understand the main job. Ask them what they are trying to accomplish. Bring this feedback into your team discussion.

Once you have the target domain and appropriate level of abstraction, formulate a main job statement following the rules outlined in the previous chapter. Start with a verb, avoid technology, and strive for stable jobs over time. Table 2.7 illustrates how to get the right scope of your main job, avoiding some of the common pitfalls.

TABLE 2.7 FORMULATE THE MAIN JOB AT THE RIGHT LEVEL

WRONG EXAMPLE	ISSUE	APPROPRIATE MAIN JOB
Select music to listen to	Stage in a big job; too small	Listen to music
Enjoy the arts	Aspirational job; too broad	Listen to music
Play music on computer	Mentions a method and a specific outcome	Listen to music
Save time by creating a list of songs	Indicates an outcome (save time) Refers to a method	Listen to music

You also want to identify related jobs and get a sense of the potential variety of goals that people have. So not only do you have to move up and down in granularity, you also need to move from side to side and recognize adjacent goals. Related jobs can help you break up a big job into more meaningful sections. In fact, some solutions (e.g., complex software programs) may address several related jobs, and it's more practical to view these individually, rather than rolled up into one.

Here are some points to consider when formulating the main job:

- **Get the phrasing right.** JTBD provides a common language for an organization, and getting the labels right is key. Iterate and refine your definition of the main job. Use a thesaurus to find the best labels. Keep it simple and one-dimensional.
- **Ensure that there is a purpose.** Main jobs should be purposeful and not actions or tasks. Strive to reflect an outcome from the individual's point of view. For instance, *look at a painting* is an action, while *understand artwork* is a JTBD with an objective.
- **Reflect an end state.** Avoid framing main jobs as ongoing activities. It's problematic to start a main job with words like *manage, maintain, keep up*, and *learn* because they don't have a clear end state. For example, *learn all there is to know in a given field* isn't

a good main job. When is learning done? Similarly, formulating a job as *manage financial portfolio* is problematic because it's hard to point to an end: managing is ongoing. Instead, phrase the job as *grow financial portfolio,* which is stronger because there is a way to be "done."

- **Separate jobs from needs.** Don't mix up needs or desired outcomes with the main job. For instance, the main job of a street vendor with a hotdog cart is to *sell food on the street.* Of course, the vendor wants to *maximize the amount of hungry people she attracts,* but that's considered a need.

Test your main job statement against these questions:

- Does the statement reflect the job performer's perspective?
- Does the job statement begin with a verb?
- Is there a beginning and end point of the goal?
- Might the job performer think, "The [object] is [verb]-ed"? (e.g., did the *financial portfolio grow*? Or was *food sold on the street*?)
- Are the statements one-dimensional without compound concepts?
- Would people have phrased the job to be done like this 50 years ago?

Getting the main job right requires some thought, discussion, and negotiation with your team. It's a fundamental decision of the scope of the customer's job and, consequently, the needs your entire business will target. Framing the main job sets your field of vision for subsequent activities. Spend time in getting the right level of granularity and formulating the job statement properly.

Define the Job Performer

Ask yourself, "Who holds the insights we need to uncover?" In some cases, it may be obvious who the primary job executor is, for instance, with simple consumer products. In B2B situations, you may need to sort out the potential roles you can target and get agreement from the team. Ultimately, you'll want to consider a range of different functions as a system of job performers, but at first the primary concern is sorting out the job performer.

Very often, the term for the job performer is directly related to the main job. Keep it simple. For instance, if the main job is *to attend a conference*, the job performer is a *conference attendee*. Or, if the main job is to *prepare a meal*, the job performer is a *meal preparer*.

Keep in mind that major factors around performing the job, such as expertise, may impact the definition of the job performer. A professional chef may go about preparing a meal differently than a home cook when preparing a meal. You can qualify the main job with relevant circumstances to get the right job performer, e.g., *prepare a meal at home*.

One approach while scoping the JTBD domain is to interview experts in a given domain initially. This often accelerates your learning about how to get a job done. So even if you frame the main job as *prepare a meal at home*, you can still learn a great deal from master chefs initially. However, unless chefs are your job performers, you shouldn't complete the JTBD research with experts. Instead, target a general group of job performers to get their insights and priorities.

Defining the main job and job performer really go hand in hand. You'll likely define both at the same time, going back and forth as you do. Interview a few potential job performers to see if you're on the right track. A few informal conversations can do wonders for narrowing in on the distinctions and labels that will make the most sense.

Form a Hypothesis About the Process and Circumstances

In a final step, start exploring the process and circumstances. You may be able to intuit some of the stages in executing the job based on existing knowledge. Try making assumptions about the sequence of stages the performer may go through. This will help guide your discussion with interviewees. But be prepared to adjust your hypothesis with new information that you'll encounter through field research.

Similarly, you may know some of the circumstances in advance. Have a conversation with your team about these to understand what you might need to probe during interviews. Make assumptions initially, but be ready to update them later after you complete your research.

Primary circumstantial factors may also influence the scope of your main job. For example, let's say you are focused on how people get breakfast. The job *get breakfast* might be too broad. If getting breakfast on a commute to work is your real focus, you can frame it as *get breakfast on the go*. Again, start broadly and qualify the main job (e.g., with *on the go*) as needed. Decide as a team what the best expression of the main job is.

Use the JTBD canvas from the previous chapter to discuss the main job, performer, circumstances, and process. Display the canvas on a large screen for everyone to see or print a poster-sized version of it to hang on the wall. This not only gets everyone thinking about a job to be done and its parts, but also helps explain the framework and labels you'll be using. Hypothesize together to start speaking the language of JTBD.

Recap

JTBD gives companies a consistent way to understand the goals and needs that customers have and then bring that insight back into the organization. It focuses on what motivates them to act: they are striving to get a job done. Think of JTBD as a type of language for teams to consistently discover and describe the goals and needs of people.

At the core of the JTBD model are five separate elements, addressing the who, what, how, why, and when/where questions of your field of inquiry:

- **Job Performer** (who): The person who will be executing the job
- **Jobs** (what): Includes a main job, related jobs, and emotional jobs and social jobs

- **Process** (how): A chronological representation of the stages in getting a job done
- **Needs** (why): The desired outcomes an individual has from performing a job
- **Circumstances** (when/where): The conditions that frame how the job gets executed

Goals are hierarchical. Through a process of laddering, you can roll up objectives to different levels. The goal at one level may be a stage for the next. In JTBD, there are four levels to consider:

- **Aspirations:** An ideal change of state, something the individual desires to become
- **Big Job:** A broader objective, typically at the level of a main job
- **Little Job:** A smaller job that corresponds roughly to stages in a big job
- **Micro-Job:** Activities that resemble tasks, but are formulated in terms of JTBD

Getting the right level of abstraction is critical. Ask "why?" to move up in the JTBD hierarchy and ask "how?" to move down.

Start your JTBD research and analysis defining the main job and job performer. Also, create a hypothesis of the job process and circumstances to be validated with research. Talk to a few people upfront for some initial insight into your overall scoping effort.

LEARN MORE

Bob Moesta, "Bob Moesta on Jobs-to-be-Done," interview by Des Traynor, *Inside Intercom.* (podcast), May 12, 2016, https://www.intercom.com/blog/podcasts/podcast-bob-moesta-on-jobs-to-be-done/

Moesta is a pioneer in JTBD and directly influenced by Clayton Christensen. In this interview with Des Traynor, co-founder of Intercom and thought leader in JTBD, he covers a range of topics. Overall, it's a good resource to understand some of the fundamentals of JTBD thinking in general.

Anthony W. Ulwick, "Turn Customer Input into Innovation," *Harvard Business Review* (January 2002).

This article was my introduction to JTBD and directly influenced my model outlined in this chapter. Ulwick essentially gives away his secrete sauce for identifying real business opportunities through needs analysis. It starts with not only a deep and thorough understanding of customer jobs, but also a consistent way of representing and working with them.

Anthony W. Ulwick and Lance A. Bettencourt, "Giving Customers a Fair Hearing," *MIT Sloan Management Review* (Spring 2008).

This article details a consistent way of expressing goals and needs, largely mirrored here. The language they provide is key for working with jobs and being able to scale them across an organization.

CHAPTER

3

Discovering Value

IN THIS CHAPTER, YOU WILL LEARN ABOUT THESE PLAYS:

- Two interviewing approaches: Jobs interviews and Switch interviews
- Analyzing insights with the Four Forces techniques
- How to map a job

I'm lucky. I regularly speak with customers. It's a privilege to observe people in their natural settings and to be able to see the world from their perspective. I thrive on it.

But few people in an organization get that chance. Think about it: How many people in your company have *never* spoken with a customer? As a result, a lot of misguided assumptions are made about what people will actually find valuable. Teams don't agree on how customer insight informs their efforts and ultimately informs growth.

It's up to you to bring market insight back to your organization. Long research reports don't work: people don't read them. If not made actionable, study findings get forgotten and have no impact.

JTBD changes things. It focuses on a clear unit of analysis: the job. This, in turn, serves as an axis for decision-making across the organization. More importantly, JTBD gives a consistent language to align around. You not only can agree that understanding customer needs are important, but also how to express and act on those needs consistently.

This isn't to say that organizations should stop other types of investigation and discovery. Surveys help understand satisfaction, usability tests improve products, and on-site visits deepen empathy. Keep doing these things.

But when done effectively, JTBD serves as a core engine of inquiry that yields long-lasting models, which can be used to drive roadmaps for years. It feeds innovation with raw materials and insights from customer discovery.

This chapter presents techniques for discovering and making sense of jobs. First, you'll have to engage people in research, primarily in the form of interview. Then you can analyze the forces of demand generation and map out the job process to make sense of your insights. Use these opportunities to get people from your organization to engage and interact with customers. The alignment of perspectives you can gain from fundamental discovery activities is invaluable later during solution-finding phases.

PLAY ➤ Conduct Jobs Interviews

Jobs don't come in neat, little packages. You have to hunt for them. You won't find jobs from analytics or marketing surveys, and you can't just "brainstorm" jobs and needs. You have to get out and talk to job performers in formal interviews.

Start by getting the right people—job performers. Then lead an open interview that lets them speak in their own words about their objectives. Don't read from a questionnaire, but instead probe on the job process and needs. Afterward, you'll need to translate what you heard into the JTBD language.

Note that jobs interviews are not intended for gaining empathy for participants per se, although that is often inevitable. Critics point out that jobs interviews miss a lot of the details about the person's overall experience. Jobs interviews also don't get at psychological states, even if there are questions about emotions and social aspects. Instead, the JTBD approach assumes that people are first and foremost motivated to get the job done so they can make progress. The interviews favor a more surgical approach to reach their goals and needs.

Recruit Participants

Jobs interviews are not research about your product or existing customers. In fact, you don't even have to talk to people who know your brand or offering. That might cloud their responses to your questions. Because you're not yet concerned about purchasing decisions or brand awareness, you just need to get job performers—the people executing your main job. It's as simple as that.

Avoid interviewing people who believe they can speak on behalf of job performers. For instance, IT procurement managers might eventually become buyers, but they typically don't carry out the job themselves. Instead, you'll want to get in contact with the potential end users for jobs interviews.

Create a screening script for recruiting that consists of criteria that you'll use to select participants. The recruiter reads off the script to find the right people. Build in stop points to disqualify a person from your research. In other words, you need to determine not only whom you want to talk to, but whom you don't want. Avoid including any reference to a specific product, service, or brand.

Recruiting scripts have three main parts.

- **Part 1: Introduction**—Include a few sentences that recruiters read word-for-word to set the stage and the right expectations.
- **Part 2: Questionnaire**—Present a series of questions to ensure that participants are job performers. Include exclusion criteria and points at which the recruiting interview can end.
- **Part 3: Schedule**—Maintain a schedule of interviews. Find a slot with qualified participants. Typically, interviews are 1–2 hours. It's better to go on site and interview face-to-face, but phone interviews are also possible

It's recommended that you use a recruiting agency. Be sure to brief the agency on the nature of the study and the interviews. If they've primarily done recruiting for marketing purposes based on demographic details, you'll have to coach them on how to get job executors. Include detailed instructions at the beginning of the screener to ensure that recruiters are clear on participants.

Otherwise, you can try to recruit job performers on your own by tapping into various sources to which you already have access. Use an online form to recruit people from your website or try recruiting through social media. Avoid recruiting only from your existing customer base to strengthen your focus on the job, not your solution.

Note that if you use existing customers for the interviews, be cautious about their bias toward your solution. You'll need to explicitly steer them away from talking about your product or solution. It's harder to interview existing customers from this perspective, but possible.

At a minimum, you'll need about 5–6 interviews to start seeing consistent patterns. It's recommended to double that number—10–12 participants are better. More participants will strengthen the thoroughness of your research. I've done studies with 20 or more interviews, adding greatly to the confidence in findings.

Incentives vary greatly depending on the domain and target participant. General consumers are willing to participate for a simple gift card worth $25 or so. Highly qualified professionals may expect hundreds of dollars for an hour or two of their time. Budget accordingly.

If you don't have a budget for incentives, recruiting will be much more difficult. Try to find some other reward, like free use of an existing product or a raffle.

Prepare for Interviews

Jobs interviews represent a type of open interviewing. It's not about reading from a questionnaire, but instead steering a conversation through specific topics. To do this, create a one- or two-page discussion guide to refer to during the session. Think of it as a list of prompts for the interviewer, not a survey for the participant.

A discussion guide typically begins with a standard greeting to set expectations. The body of the discussion guide consists of a list of questions around the relevant themes. These prompts should address the questions, assumptions, and gaps in knowledge you have. It's rare you'll cover the themes in the same order as in your guide during a conversation, but that's OK. Go with the flow and use the discussion guide to get back on track as needed.

It's a best practice to do the interviews with more than one researcher. One is the primary interviewer, and the other acts as an observer. Maintain these roles. This focus allows the lead researcher to build a rapport with the participant and steer the conversation. The observer may ask questions at the end or when asked.

Conduct Interviews

People don't know how to create solutions to solve their problems, so don't ask them. They do know about their own objectives and needs. Focus on understanding their jobs to be done from their perspective.

Allow the conversation to unfold naturally, but use the discussion guide to bring the participant back on topic as needed.

After greeting the participants and setting the stage, begin by asking the participants about themselves and their general situation related to the main job. Remember that you are not there to get product feedback. If they want to ask you specific questions about a specific solution, hold that until the very end. Then go back and address it after the interview.

Make sure that the interview is about them and their jobs. Strive to do more listening than talking. Use a combination to probe on the job, the process, and their needs and emotions.

1. **Get background about the participant and the job.**

   Build rapport and get the participant talking freely.

   - Tell me a little about yourself and what you do.
   - When was the last time you did the main job?
   - How did you feel overall while getting that job done?

2. **Understand the main job and related jobs.**

   To get participants to talk about their jobs to be done, ask questions like the following:

   - What are you trying to accomplish? What tasks are involved?
   - What problems are you trying to prevent or resolve?
   - What helps you achieve your goals?
   - What would the ideal service be to do the job for you?
   - What else are you trying to get done?

3. **Understand the process of executing the job.**

   Go through the stages of getting the job done.

   - How do you get started?
   - What is the previous step? What's the next step?

- How do you continue after that?
- How do you make decisions along the way?
- How do you feel at each stage in the process?
- How do you know you are doing the job right?
- How do you wrap things up?

4. **Find needs.**

   Uncover the desired outcomes that people are looking for while performing the job.

   - What workarounds exist in your process?
   - What do you dread doing? What do you avoid? Why?
   - What could be easier? Why?
   - Why do you avoid doing certain parts of the job?
   - What's the most annoying part? Why is that frustrating?
   - How do you feel when the job is completed?

5. **Probe on circumstances.**

   Find out when and where performing the job makes a difference. Try to uncover the most salient factors that frame getting the job done.

   - In which situations do you act differently?
   - What conditions influence your decisions?
   - How do the environment and setting affect your attitude and feelings while getting the job done?

While interviewing, steer the granularity of answers. Ask "why?" to get more general and move toward outcomes. Ask "how?" to encourage the interviewee to be more specific and to dig into the process.

Try also sketching a diagram of the process together with the participant as the interview goes along. Point to specific steps in the sketch to clarify the process. Then dig deeper to understand their goals and feelings at each step, along with the context of getting the job done.

People tend to generalize when responding to open-ended questions, and they may speak broadly about their thoughts and behaviors. To get more specific and keep the conversation about their experiences, use the *critical incident* technique. There are three simple steps to follow.

1. **Recall a specific incident.** Have them remember a time when executing the job went particularly wrong.
2. **Describe the experience.** Ask them to describe what happened, what went wrong and why, and how they felt at the time.
3. **Discuss the ideal state.** Finally, ask what should have happened and what would have been ideal. This helps reveal their underlying needs.

Keep track of time during the interview and be respectful of participants' time. Wrap up with a brief conclusion, thanking them and letting them know the next steps, if any.

Follow these general interviewing tips and guidelines:

- **Create rapport.** Use eye contact, nod, and agree (e.g., "Yes, I see how that's frustrating.").
- **Listen.** Let them do the speaking. Don't put words in their mouth.
- **Avoid yes-or-no questions.** Ask open-ended questions to keep participants talking.
- **Dig deep.** Follow interesting thoughts (e.g., "Tell me more about that.").
- **Minimize distractions.** Avoid interruptions and getting off track. Stay focused.
- **Go with the flow.** Make the best of suboptimal situations.

- **Don't interrogate.** Create a comfortable interaction, like talking to a friend.
- **Use pauses.** Breaks give the interviewee a chance to think and respond.
- **Research in pairs.** Designate one person as the interviewer and the other as the notetaker.

Analyze the Data

Schedule time to debrief immediately after each session or two. Review notes with your interview partner. Take the time to complement each other's understanding of what the participant said. If you wait too long to review your notes, you may forget details and lose the context.

Typically, your notes will suffice for capturing data, but you should also record the audio of the interviews for a backup. Ideally, you'll go back and listen to the recordings as you do analysis, but that often proves too time consuming. If you want to be extra thorough, have each recording transcribed word for word. Keep in mind that a 60-minute interview will generate about 30 pages of text.

For one project I conducted, the team coded JTBD insights in real time. There was one interviewer and one notetaker. The notetaker recorded desired outcome statements as heard during the interviews. The result was that we arrived at a usable data set quickly after each interview without having to analyze our notes or recordings.

Create a spreadsheet to extract relevant observations. Put direct observations and quotes in the first column. Then create four columns for interpretations: micro-jobs, emotional and social aspects, needs, and circumstances. Translate the observation into the JTBD language following the rules of formulation, as shown in Figure 3.1.

OBSERVATIONS	INTERPRETATIONS	
Quotes and notes	**Job steps**	**Emotion/Social Aspects**
PARTICIPANT #1		
"There are so many conferences these days, it's hard to choose which ones to attend. I don't even know the ones I don't know about. How can I decide?"	Decide which conference to attend	Feel overwhelmed with available options
This participant mentioned that her company only pays for employees to go to one conference a year and she has to make a case to her boss.	Convice manager to attend	
"I like to go back to conferences I've attended in the past because I know people and I know how the event will go, what the basic flow is, you know? I don't have to figure out what to do or where to get started as much as completely new conferences."	Orient to the conference	Feel a sense of familiarity
We observed this participant and others taking photos of key slides during an event instead of taking notes.	Capture conference content	
This participant expressed a desire to have a good deal of networking opportunities during the conference.	Meet people; talk with other conference goers	
"I like the community aspect too."		Be seen as a member of a professional community
The participant complained about conference wifi. It's often spotty, slow, or doesn't work at all.	Communicate with others outside of the event	

Strive to start with actual quotes from the interviews. Otherwise, paraphrasing works well. The elements of JTBD discussed in the previous chapter become the filter by which you'll organize insights.

- **Job steps:** Indicate steps in getting a job done and the micro-jobs you find during the interview. Be sure to begin each with a verb and omit any reference to technologies or solutions.
- **Emotional and social aspects:** Record emotional aspects beginning with "feel" or "avoid feeling" and then social aspects with "be perceived as" or "avoid being seen as."
- **Needs:** Listen for answers to your "why" questions, as well as hacks, workarounds, avoidances, and procrastinations. Be sure to note needs beginning with a verb that shows the direction of change.
- **Circumstances:** Note any situational constraints beginning with "when."

INTERPRETATIONS		COMMENTS
Needs	**Circumstances**	
Minimize the effort it takes to select the right conference to attend		
	When attendance is limited	
Miminze uncertainty around the conference program; maximize familiarity with the conference	When it's a repeat conference	
Minimize the amount of note-taking		See photos from the conferences
Maximize the number of network contacts at the event		

FIGURE 3.1
Analyze your raw notes by reformulating observations into JTBD terms.

Refining and rewriting job statements is important. The form of your insights matters in JTBD. Pay attention to formatting and the rules of job statements. Do your statements represent stable intent over time? Are they devoid of technology, solutions, and methods? The goal is to create a master list of job steps, needs, and circumstances across all of your interviews.

It's also possible to validate your insights with participants. For instance, you can schedule a second round of interviews to get feedback on your need statements. Don't necessarily show participants statements or read them, but instead interview them about their jobs and touch on the themes of the needs. Pay attention to whether the need statement resonates and applies to them or not. See the case study later in this chapter for more details on validating job statements.

LEARN MORE ABOUT THIS PLAY:
CONDUCT JOBS INTERVIEWS

Steve Portigal, *Interviewing Users* (NY: Rosenfeld Media, 2013).

This book is one of the best and most complete volumes on interviewing. It's very practical and hands-on, including a range of sample documents and materials to refer to online. Portigal covers the interviewing process from end to end, including setting objectives, recruiting, interviewing, and analysis.

Giff Constable, *Talking to Humans* (Self-published, 2014).

This thin volume of only 75 pages provides an excellent overview for getting in front of people and talking to them. There is a wealth of practical information for getting started and conducting quick interviews.

Mike Boysen, "A Framework of Questions for Jobs to Be Done Interviews," *Medium* (blog), 2018.

Mike is a practitioner of ODI at Strategyn, and his guide to jobs interviews is one of the most complete to date specifically focused on JTBD. He provides an overview of jobs interviews, as well as a detailed worksheet for dissecting interview data. Although there are many similarities, the approach I present in this chapter differs from how Strategyn conducts jobs interviews in practice.

Hugh Beyer and Karen Holtzblatt, *Contextual Design* (San Francisco: Morgan Kaufmann, 1998).

This landmark book in software design brought us the concept of contextual inquiry, or research in the context in which users work. The authors present a comprehensive process for finding and distilling insights and then tying them directly to software modelling and design. Their notion of focusing on users' "work" discussed throughout the book overlaps greatly with JTBD.

PLAY ➤ Run Switch Interviews

An alternative form of interview to a jobs interview is the so-called Switch interview. JTBD typically focuses on either jobs interviews or Switch interviews as a main style of research. It's also possible to mix the two approaches.

The Switch technique was developed and made popular by Bob Moesta and Chris Spiek to answer the question, "Why do customers 'hire' a given product?" The idea is to reverse engineer why people switch from one way of doing a job to another in order to uncover their underlying intent.

The approach seeks to recreate the purchase journey, starting with a concrete product in mind. When the purchaser is also the job performer, as in consumer contexts, there is insight into the reasons for executing a job as well. But in B2B situations, the purchaser may be separate from the job performer. It takes a lot of practice in recognizing patterns of needs to pick up on job performer outcomes when speaking with a buyer in a B2B situation, but it's possible.

The Switch Timeline

A simple timeline is used during the interview instead of a discussion guide. Because it may be hard for participants to remember their first thought—or their original needs—the technique works backward through the points on the timeline. Get participants to give you the "documentary" version of their purchasing journey.

The timeline represents the search for a solution and has phases that are articulated with specific events or moments in time. There are six phases to work through, each articulated by a key event, as seen in Figure 3.2.

- **First thought:** This is the initial moment that a change is needed, often implicit.
- **Passively looking:** The buyer is not putting energy into a search, but notices options. The first event makes the search explicit.

- **Actively looking:** The buyer invests time and energy into seeking a solution. The second event transitions the buyer into a purchase decision.
- **Deciding:** Here, the buyer consciously weighs alternatives. This phase ends with a decision to buy.
- **Consuming:** After making a purchase, the buyer uses the product or service. Either the product experience is completed, or it's ongoing.
- **Satisfaction:** The solution either leads to progress, or it doesn't.

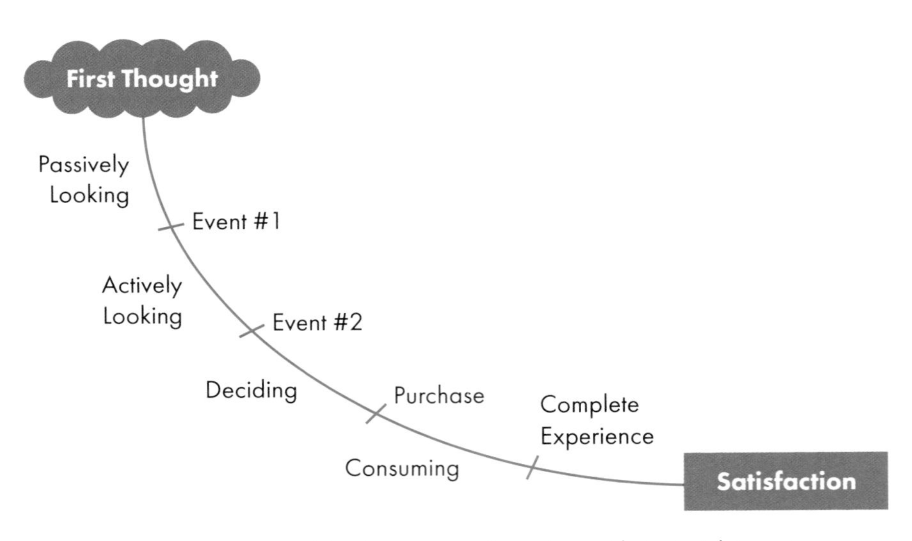

FIGURE 3.2 Use a standard timeline as your discussion guide in Switch interviews.

Use a timeline like the one shown in Figure 3.2 to guide your discussion and record notes. Work backward as needed, digging deeper as you go. Ask, "What happened before that?" along with "Why did you make that decision?" Try to find the motivation or the energy, as Moesta puts it, behind the switch from old to new.

In many ways, Switch interviews are similar to the critical incident technique, noted in the previous JTBD play. The incident, in this case, is the purchase. You're using that as a specific event in the past to focus your conversation.

The goal of Switch interviews is to find the underlying motivation around buying and using a product. The intent is to understand the "switch" behavior. In an interview with Des Traynor, founder and Chief Strategy Officer of Intercom, Moesta talks about finding the energy in the ultimate JTBD:[1]

> If I walk in with the premise that something caused someone to do something and then buy something, the product or the service is only part of the solution, so it's almost getting elevation above it. You can start to see the dominoes, those little causal things that have to click into place to happen. We're not really interested in what people are just saying. It's what they're doing or not doing, and almost the energy that's going through it.

From this standpoint, the switch doesn't have to be between products of the same kind. It's really about changing from an old way of working to a new way.

Variations on Switch

Switch interviews come from a marketing perspective and frequently focus on the purchase of physical consumer products. It's much more complicated to understand the first thought for buying an online software service in a B2B context. For one thing, in B2B situations, the buyer and job performer may be two separate people. But also there is often no emotional connection.

Working for Intercom, a leading customer messaging platform for online software providers, Sian Townsend modified Switch interviews slightly

1. See Des Traynor's interview with Moesta: "Bob Moesta on Jobs-to-be-Done," *Inside Intercom* (podcast), May 12, 2016, https://www.intercom.com/blog/podcasts/podcast-bob-moesta-on-jobs-to-be-done/

for her needs. She and her team found that buying software for a company to use is different than buying, say, a mattress or physical product.

Using the basic Switch technique as a framework, they came up with a slightly different approach. To be able to interview users of their B2B solution, they recalibrated the questioning to be more appropriate for complex software. Their key was to focus less on the buying decision as an event and more on the Switch from one product to another, as well as shifting from emotional triggers to more functional motivations.

She and her team asked questions like:

- What was working at the company in the past?
- What tools were you using before you got the new software?
- Were you involved in buying the new software?
- Who decided to move to a new software solution?
- Who made the decision to switch?
- Are they still in the company now?
- What impact does the new software have on your work now?

The point is to jog someone's memory to recall why things got so bad that the company decided to switch. Then focus on the impact on their work after the switch.

Customer Case Research

Switch interviews are based on a technique called *customer case research* (CCR), first documented by marketing experts Denise Nitterhouse and Gerald Berstell in 1997.[2] Nitterhouse and her team had been practicing the technique for a decade earlier, so the method dates back to the late 1980s.

Like the Switch interview, CCR is centered on in-depth, qualitative interviews to uncover insights missed by other forms of research.

2. Gerald Berstell and Denise Nitterhouse, "Looking 'Outside the Box': Customer Cases Help Researchers Predict the Unpredictable," *Marketing Research* 9 (1997): 5.

Opportunities tend to fall into one of seven common categories they've discovered after thousands of interviews:

- **Unexpected openings** are situations where customer behavior is a total surprise and not identified with other marketing research techniques, such as surveys and focus groups.
- **Embedded segments** reflect multiple ways of buying the same product and are uncovered with CCR interviews.
- **Unanticipated decision criteria** also emerge from qualitative, on-site interviews with customers.
- **Hidden decision-makers** can be identified with CCR, challenging the assumptions of traditional methods as to who makes decisions ultimately.
- **Unintended product uses** are typically uncovered in CCR, pointing to alternative means of getting a job done.
- **Unforeseen obstacles of acceptance** are criteria for gaining market adoption of an offering, but often are missed by traditional marketing methods.
- **Unarticulated needs** emerge from CCR interviews, bringing opportunities for new product and service concepts to be developed.

As you can see, CCR is not only inline with JTBD thinking in general, but it's also more robust than Switch interviews alone. Additionally, target participants include customers who have recently switched solutions, which the authors call "switchers." But CCR also considers other perspectives and types of customers, including polygamists (customers with multiple vendors for the same product), newbies, quitters, and persisters.

CCR is a technique that is firmly grounded in marketing, primarily used to increase sales. It hasn't made its way into product and service innovation, although insight can be useful for developing a new offering as well.

In the end, the specific research technique that you use—jobs interviews, Switch, CCR, or something else—depends on your goals. If you want to initiate a JTBD effort for a brand new solution, you might be better off conducting jobs interviews. If you already have a product and an established market, then you may want to start with your offering. It depends on if you want to be more abstract and focus on the job independent of an existing market, or if you want to take your solution as a starting point and deduce jobs from there.

LEARN MORE ABOUT THIS PLAY: RUN SWITCH INTERVIEWS

Chris Spiek and Bob Moesta, *Jobs-to-Be-Done: The Handbook* (Re-Wired Group, 2014).

> This thin volume focuses on interviewing around a timeline. There are practical tips for conducting interviews and eliciting key information. In particular, the authors focus on the forces of progress toward a new behavior: the push and pull factors involved in a new choice, and the forces that block change. This is a practical guide that includes a wealth of tips and recommendations for conducting interviews in the field. For an example of a Switch interview, see the popular "Mattress Interview" on the JTBD Radio website: http://jobstobedone.org/radio/the-mattress-interview-part-one/

Gerald Berstell and Denise Nitterhouse, "Looking 'Outside the Box': Customer Cases Help Researchers Predict the Unpredictable," *Marketing Research* 9 (1997): 5.

> Customer case research (CCR) is the clear precursor to Switch, representing the same type of exploratory research to discover critical purchase drivers. Like Switch, CCR uses in-depth interviews to trace back to the moment when people first decided to switch solutions. CCR begins without preconceptions about purchase criteria, allowing for unintended uses and unanticipated decision criteria to emerge.

PLAY ➤ Analyze the Four Forces of Progress

Bob Moesta and his team at Re-Wired have demonstrated that there are four forces that drive behavior of switching from one offering to another. To analyze your interviews and pinpoint insights, they developed the four forces diagram, pictured in Figure 3.3.

At the center are the four forces of change:

- A **problem** with the current product leads a consumer to consider a new solution.
- The **attraction** of a new product pulls them away from their existing ways of working.
- **Uncertainty** about change provides a reason to stay.
- **Habits** keep consumers from switching.

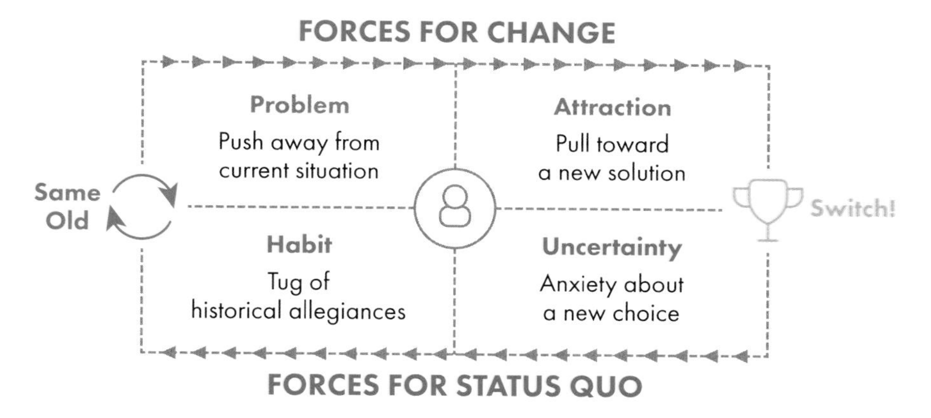

FIGURE 3.3 Use the Four Forces model to understand the reasons for switching and staying.[3]

Moving from left to right on the top row, we see the dynamics of push and pull factors that move customers toward choosing a new behavior.

3. Thanks to Brian Rhea for use of this version of the Four Forces diagram.

The bottom row shows the forces holding them back, or the blockers of change. In a nutshell, if the push of an existing problem and the pull of a new solution (top of diagram in Figure 3.3) are greater than the uncertainty of switching and the habit of a current solution (bottom of diagram in Figure 3.3), then a person will switch from one behavior to another. If not, inertia sets in, and there is no change.

Four Forces analysis is really about innovating demand generation shaped by push and pull factors. Alan Klement, for one, believes that customers pull offerings to reach more aspirational objectives. In an article entitled "The Forces of Progress,"[4] Klement writes: "People don't buy products just to have or use them; they buy products to help make their lives better (i.e., make progress)."

Magnifying the dynamics of the forces a bit, we can get more specific about both push and pull factors. On the push side, there are both internal and external factors to consider. Sometimes external conditions in the environment push people away from an existing solution, such as company reorgs, new laws, or life changes (e.g., getting married, having a baby, etc.). Internal factors reflect a change in attitude or belief. For instance, a person might come to the realization they are unhealthy and want to start a diet or new workout program after being motivated to do so.

Solutions themselves have a pull that draws consumers in with a preference. Sometimes demand can even be created. For example, no one needed a smartphone until everyone needed a smartphone.

Four Forces in Action

The steps to completing a Four Forces analysis are straightforward:

STEP 1 ➤ Conduct research.

Using the Four Forces model naturally follows the Switch interview technique. Be sure to include questions that probe at each of the four forces.

4. Alan Klement, "The Forces of Progress," *JTBD.info* (blog), May 17, 2017, https://jtbd.info/the-forces-of-progress-4408bf995153

For instance, once you get to the first thought, probe on what problem people had that was worth solving, what attracted them to new solutions, when uncertainties arose, and what habits they had to overcome.

STEP 2 ➤ Extract insights around each of the forces.

You can use the diagram to organize your research findings. Divide a sheet of paper or document into four quadrants and label them with "problems," "attraction," "anxiety," and "habit" going clockwise from the upper left. Then, as you review your notes from your interviews, sort your insights into each of the categories. When completed, you should have an outline of the reasons for switching. Do this across multiple interviews until clear patterns start to emerge.

The Re-Wired Group makes two templates for conducting Switch interviews and making the Forces analysis available, shown in Figure 3.4.[5]

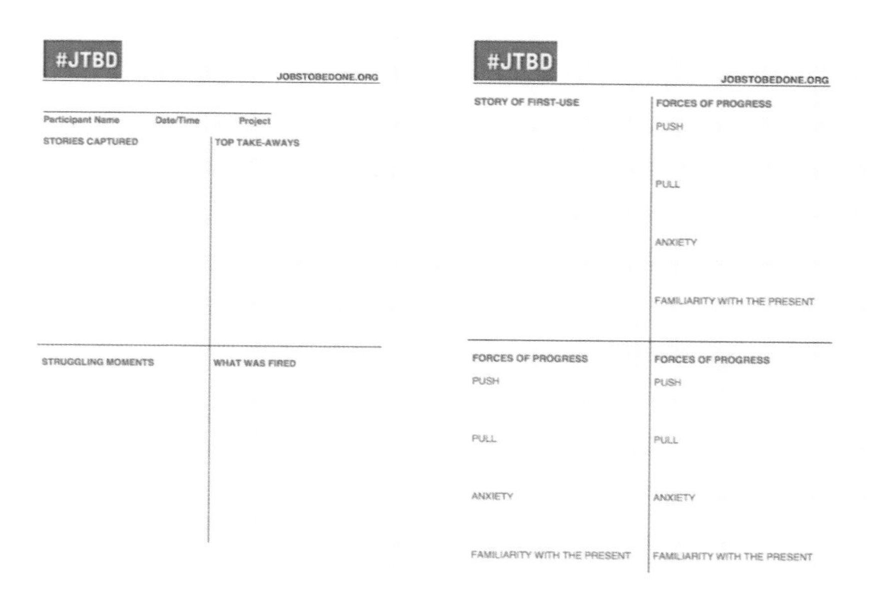

FIGURE 3.4 A simple worksheet helps analyze data from Switch interviews.

5. See the resources on www.jobstobedone.org

Try debriefing after several interviews with your team. Each person should tell the stories of their respective timelines. As one person talks, the others fill out the Four Forces model. Set a timer for each person to move through a series of interviews quickly. At the end, you'll have a documented, shared understanding of the reasons why people switch from one way to another.

STEP 3 ➤ Find your opportunity.

You're ultimately looking for the moment when people start struggling. At that point, there is usually a job to be done. Once you identify the job, you can then record it using the rules of formulation discussed in the previous chapter. Note emotional, social, and related jobs as well. In addition, try mapping the job discussed in the next play in this chapter.

Finding a way to address people's struggles is the seed for innovation. The diagram points to opportunities by outlining where the struggles lie. Is it in the push or in the pull factors? Is the old way holding people back more than the new way is attracting them?

Note that rarely is just one of the forces in play, but rather a combination of multiple forces at the same time. Your plan after conducting a Four Forces analysis will likely include action items around each of the dimensions.

For instance, the provider of an online travel expense tool may find problems with existing methods: collecting, organizing, and submitting receipts is very time consuming. An obvious step is to highlight the pain the solution addresses (*problem*), as well as the ease of submitting receipts from a mobile phone while on the go (*attraction*).

But in order for the application to be adopted, finance managers will need to follow a new process. They may wonder if employees will submit expenses correctly? Will the new system be able to scan receipts as promised? Analysis with Four Forces can show the specific *uncertainties* and *habits* that finance managers may have, which can be addressed in training and education materials.

Extending the Forces Diagram

The Four Forces diagram is a way to synthesize insight from qualitative interviews. It's typically used to create new solutions or improve existing ones. But its application extends to any situation in which change is needed.

Kevin Kupillas, design manager at HubSpot, shows how the Forces diagram can be used in other settings as well and include some of the following extended applications of the technique:[6]

1. **Interviewing job candidates.**

   Use the Four Forces to understand the struggles of a candidate's current job and how you might get them to join your team. Frame questions around each of the forces.

   - **Problem (push):** What do you struggle with at your current job? What's less than ideal?
   - **Attraction (pull):** What excites you about the new role or company? How do you imagine your life improving?
   - **Uncertainty (anxieties):** What concerns do you have about the new role or company?
   - **Habit (familiarity):** Is there anything you would miss about your current job?

2. **Introducing change in a company.**

   When you get to the bottom of it, the forces are really about any type of change. Use the technique to find where the biggest hurdles may lie. For instance, if a company is rolling out Agile

6. Kevin C. Kupillas, "May the Forces Diagram Be with You, Always," *JTBD.info* (blog), September 21, 2017, https://jtbd.info/may-the-forces-diagram-be-with-you-always-applying-jtbd-everywhere-b1b325b50df3

practices to the product development team, you can use the forces to diagnosis potential fail points:

- **Problem (push):** What are the struggles of today? What motivates the team to change?
- **Attraction (pull):** What do people look forward to with the new way of working?
- **Uncertainty (anxieties):** What are people most uncertain about from adopting change?
- **Habit (familiarity):** What existing practices do people have to change?

In the case of an organization adopting Agile, it may be obvious that there is a problem with their current way of developing software: projects are over time and over budget, and the solution doesn't meet user needs. Agile is attractive because it offers flexibility and the ability to pivot midway. But if everyone isn't comfortable with a change, there will be some anxiety around adopting Agile—for instance, learning new tools and practices while trying to develop software. And even with the best training, people may fall back on old ways of working.

3. **Making a personal change.**

We all struggle with changing our bad habits—whether it's sticking to a diet or staying in touch with family and friends. Or, in Kupillas' case, convincing friends to write more on *Medium*.

- **Problem (push):** What pains do people have today when changing a habit?
- **Attraction (pull):** What are the benefits? What will you get out of it?

- **Uncertainty (anxieties):** What could go wrong? What are people apprehensive about?
- **Habit (familiarity):** What makes people lack action? What habits hold them back?

Clearly, the Four Forces are not just about switching products, but also about changing from one state or way of working to another. In the context of your business, the Four Forces ultimately show product-market fit, or how well your offering meets the needs of a given market. Use the technique to diagnose the key aspects in generating more demand with your solution.

LEARN MORE ABOUT THIS PLAY:

ANALYZE THE FOUR FORCES OF PROGRESS

Alan Klement, *When Coffee and Kale Compete* (Self-published, 2016).

This self-published book from Klement, an outspoken voice in JTBD, contains a wealth of practical information for applying Switch techniques, including the Four Forces analysis. Klement focuses a great deal on demand generation, crossing the line between product development and marketing. Klement is a proponent of the "jobs as progress" view of JTBD in general, reflected throughout this book. Also see his article "The Forces of Progress" on the JTBD.info website (2017).

Chris Spiek and Bob Moesta, "Unpacking the Progress Making Forces Diagram," *JTBD Radio* (podcast), February 23, 2012, http://jobstobedone.org/radio/unpacking-the-progress-making-forces-diagram/

In this podcast, the creators of the Four Forces diagram discuss its intent and use in detail. You can either listen to the podcast or read the transcript. The material is conversational and, as a result, very accessible. Spiek and Moesta give a lot of examples to illustrate their points as well.

PLAY ➤ Map the Main Job

The main job is a process that you can map chronologically. From your interviews, create a sequence of stages in a visual representation that show underlying patterns of intent. What are the subgoals a person has while performing a job? What are the phases of intent that unfold as the job gets done?

The aim is to organize your research and discover how the process unfolds. Note that it's not about mapping tasks or physical activities, but about creating a sequence of smaller goals that make up the main job. Ideally, the job map will not include any means of performing the job.

A job map is not a customer journey, service blueprint, or workflow diagram. It does not reflect what a person does to discover, learn about, select, buy, and use a product or service. These activities are relevant to the buyer and purchasing process, which is treated separately later. Instead, the map reveals the process of completing the job from the executor's point of view, not the buyer or customer perspective.

Tony Ulwick introduced the concept of a job map as part of his Outcome-Driven Innovation (ODI) method. The intent is to illustrate what the job performer is striving to get done at each stage in executing a job. Together with his colleague, Lance Bettencourt, the team suggests a universal structure for all job processes with eight standard phases (see Figure 3.5).[7]

1. **Define:** Determine objectives and plan how to get the job done.
2. **Locate:** Gather materials and information needed to do the job.
3. **Prepare:** Organize materials and create the right setup.
4. **Confirm:** Ensure that everything is ready to perform the job.
5. **Execute:** Perform the job as planned.

7. Lance Bettencourt and Anthony W. Ulwick, "The Customer-Centered Innovation Map," *Harvard Business Review* (May 2008).

6. **Monitor:** Evaluate success as the job is executed.
7. **Modify:** Modify and iterate as necessary.
8. **Conclude:** End the job and follow-up.

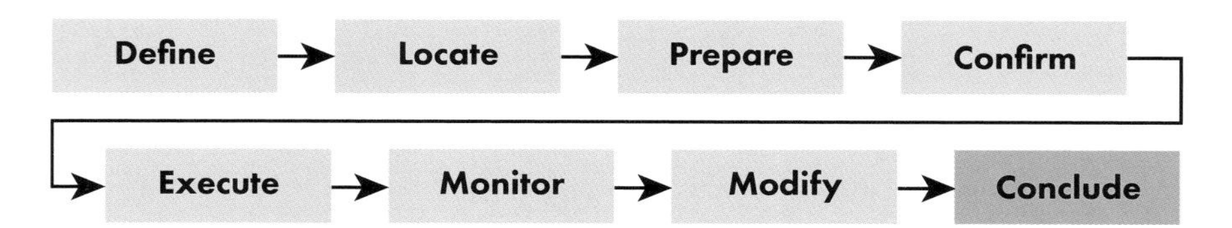

FIGURE 3.5 A universal structure of a main job has eight stages.

Consider these stages as more of a checklist than a prescriptive model. The point is to remember to cover all types of stages involved in executing the main job—before, during, and after. You'll have to modify their labels as needed to describe your particular main job. Keep the labels short, ideally expressed as single-word verbs. The list below reflects some common verbs for each of the stage types in the universal job map.

1. Define, Plan, Select, Determine
2. Locate, Gather, Access, Retrieve
3. Prepare, Set up, Organize, Examine
4. Confirm, Validate, Prioritize, Decide
5. Execute, Perform, Transact, Administer
6. Monitor, Verify, Track, Check
7. Modify, Update, Adjust, Maintain
8. Conclude, Store, Finish, Close

Each stage should have a purpose and be formulated as a functional job. Avoid including emotional and social aspects in the stage labels, and avoid bringing in adjectives and qualifiers that indicate a need, like "quickly" or "accurately." Strive to make the stages as universal and stable as possible without reference to the means of execution. Jobs are separate from solutions.

STEP 1 ➤ Create a job map.

Since job maps are chronological, it's easier to start with the three large phases of the main job: beginning, middle, and end. Place the labels for each category on a document and arrange the micro-jobs uncovered from your interviews in the appropriate category.

To demonstrate, let's assume that you're looking at the main job to *facilitate a workshop* and the job performer is a *workshop facilitator*. Figure 3.6 shows how you can start clustering findings from your research into the three large categories.

Then continue to group the jobs into about eight stages (see Figure 3.7). Use the universal stages as a starting point, but change the labels as needed. Language is important, so spend time refining the labels and divisions as you go. The important thing to remember is to map the

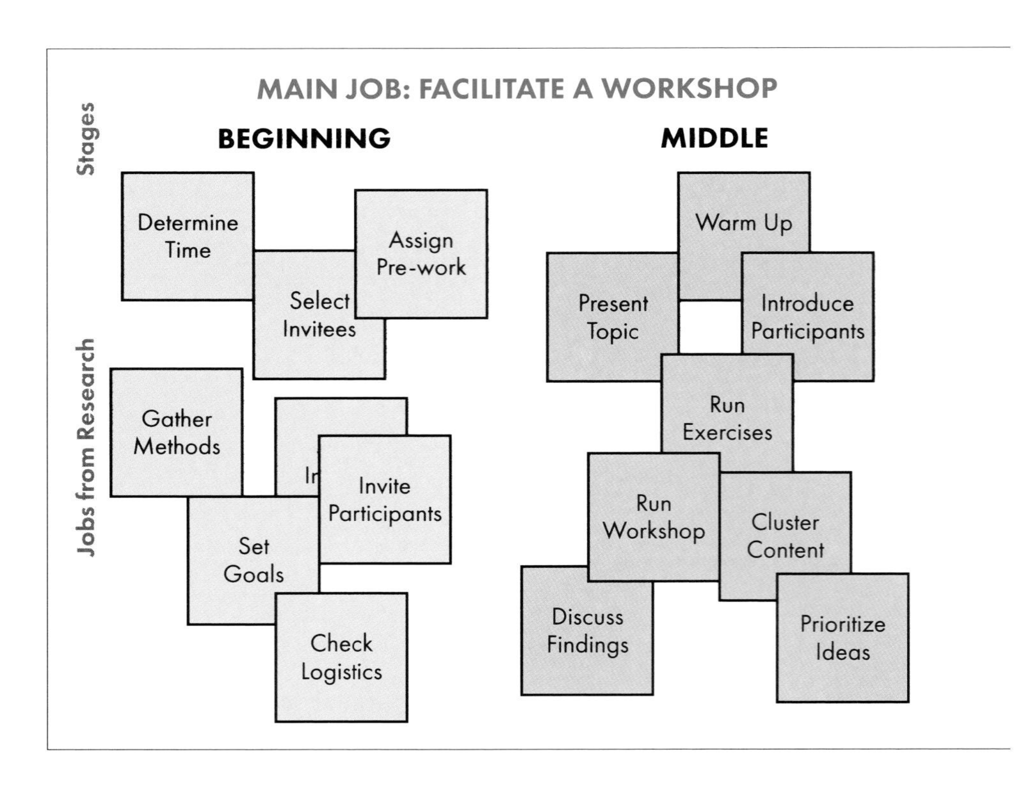

entire job, from beginning to end, from the job performer's point of view. The map is not about buying a product or interaction with a brand; it's about how the job performer gets the job done.

You may end up with fewer or more stages. It's also possible to include a loop for iteration or even a branch in the flow. The resulting diagram should stand as a clear model for describing the process of performing the job that everyone in your organization can relate to.

Ideally, you'll validate the overall model with job performers. After you complete a first draft of the map, talk through it with some of the interview participants. If the labels and the divisions between stages need a great deal of explanation or seem to be confusing, rework them until it's simple enough to be self-evident.

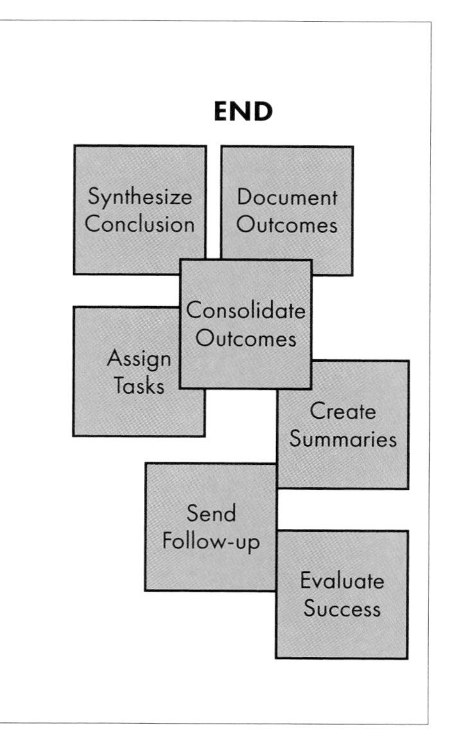

FIGURE 3.6 Arrange smaller jobs uncovered in interviews into three categories: beginning, middle, and end.

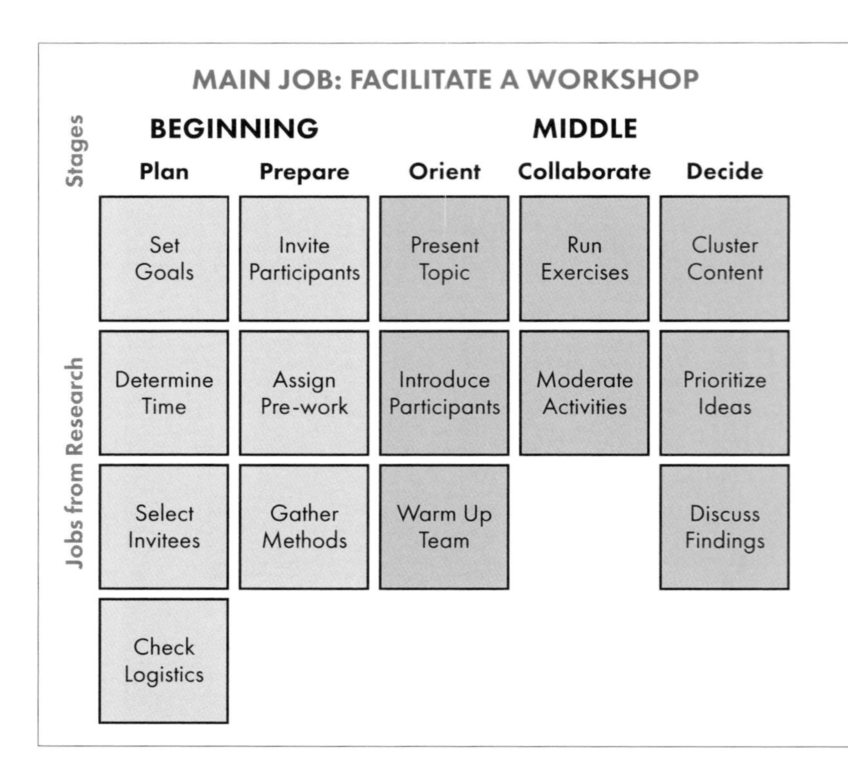

STEP 2 ➤ Put the job map to use.

With a job map in hand, organizations can create better products and services that people actually need. Bettencourt and Ulwick urge teams to use job maps collaboratively to identify opportunities. They write:[8]

> You can begin to look systematically for opportunities to create value... A great way to begin is to consider the biggest drawbacks of current solutions at each step in the map—in particular, drawbacks related to speed of execution, variability, and the quality of output. To increase the effectiveness of this approach, invite a diverse team of experts—marketing, design, engineering, and even some lead customers—to participate in this discussion.

8. Ibid.

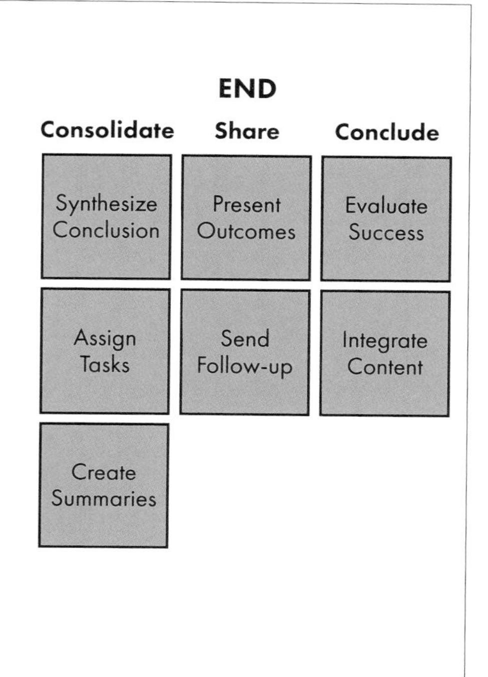

FIGURE 3.7 Determine the stages of the main job by clustering micro-jobs found during research.

Innovation opportunities can come at any stage in the job map. Consider these examples:

- Weight Watchers streamlines the "Define" stage with a system that does not require calorie counting.
- To gather items during the "Locate" step while moving houses, U-Haul provides customers with kits that include different types of boxes needed.
- Nike helps joggers evaluate the success of the job in the "Monitor" step with a sensor in the running shoe that provides feedback about time, distance, pace, and calories burned via a connection to an iPod.
- Browser-based SaaS software updates automatically so that users don't have to install new versions, thereby reducing complexity in the "Modify" step.

The job map helps find opportunities for your business systematically, as well as ways to create new value. Once completed, ask yourself these and other questions to get started:

- Is there a more efficient order of stages in performing the job?
- Where do people struggle the most to get the job done?
- What causes the job to get off track?
- Can you eliminate stages or steps along the way?
- How might the job be carried out in the future, given current trends?
- How might you get more of the job done for customers?
- What related jobs can your offering address or tie in to the job?

The job map ultimately defines the scope of your business. Align your solutions to it to spot gaps and opportunities. Compare alternative offerings and means of getting a job done for competitive insight. Prioritize areas within the job process to drive your service roadmap. Find opportunities that can be reflected in marketing campaigns and sales pitches.

LEARN MORE ABOUT THIS PLAY: **MAP THE MAIN JOB**

Lance Bettencourt and Anthony W. Ulwick, "The Customer-Centered Innovation Map," *Harvard Business Review* (May 2008).

> This article discusses the job mapping process in detail with clear guidance on how to formulate the phases. The authors describe the principles behind job mapping, as well as how to spot opportunities.

Jim Kalbach, "Experience Maps," Chap. 11 in *Mapping Experiences* (Sebastopol, CA: O'Reilly, 2016).

> In this chapter, I discuss job maps briefly in the context of experience maps. Also included are highlights from similar diagramming techniques.

Combining JTBD Switch Interview with Concept Testing

By Steph Troeth, UX Strategist/Researcher

A QUESTION OF TRUST

While I was the head of research at Clearleft, we were tasked by the digital team at Greater London Authority to design and undertake a research program. Our goal was to ensure that London.gov.uk continued to meet the needs of its audience. One of the projects that required urgent user research was Talk London, an online community platform that enables London's citizens to participate in the city's policy-making process.

The digital team had embarked on a program to integrate a number of microsites onto the main London city government's platform. However, until that point, Talk London had been an independent, safe space for citizens to engage with policy. It was important to discern if the integration with the main London.gov.uk's website—which had strong links with the Mayor's identity—would affect users' understanding of the site's purpose and value. We would also want to ensure that their sense of trust was not compromised, and that they would feel confident to contribute to consultations, surveys, and discussions.

METHODOLOGY

We set up hour-long interviews with five existing users who were active on the forum and five users who had never heard of Talk London but demonstrated propensity for engagement on London issues in other ways.

I saw an opportunity to apply the JTBD Switch interview in this context because past experience had shown me that it was particularly effective in surfacing emotional and social jobs. By setting ourselves up to understand the jobs that users hoped to achieve with Talk London, we could better understand how we should (or shouldn't) undertake the integration.

CONTINUES ➤

CONTINUED ➤

Given that the platform integration also meant a change in information architecture and navigation, it was important that we ensured the new designs were well understood and usable. On the original website, it was not obvious which surveys or forum threads were currently active. Working with the team, we decided that we could also take the opportunity to explore how we could better surface the ways that users could engage with Talk London through designing two different prototypes.

I used the Switch interview method somewhat stealthily as the focus of the research on the evaluation of the concepts and not directly on the jobs of the target audience. But in surfacing the emotional and social jobs in the opening interviews, I hoped to be able to get a truer evaluation of the proposed designs.

ADAPTING THE JTBD SWITCH INTERVIEW

The usual way to kickstart a Switch interview is to begin at the "first thought," or at a point of commitment or purchase. For Talk Londoners, we began at the point of commitment (when they signed up) and worked backward, diving deep on how they ended up using the forum and their patterns of engagement. We looked to understand their core jobs and the nuances around these needs. For non-Talk Londoners, we began as closely as possible to a "first thought" and focused on how they were currently engaged on a specific issue to see if their core jobs were similar to that of Talk Londoners. We were particularly interested in what might present as barriers to engagement. We then showed all participants the existing website and the new design concepts.

Interestingly, what we found was that the most important thing to everyone (the "job," if you will) was the sense of having an impact through their contributions. They were willing to give up time to *contribute with the hope of having impact,* despite not being sure if they had done so with their actions (see Figure 3.8). Participants were savvy; they understood that policy issues were complicated and could take a long time. They believed that Talk London could have an actual impact—or it could just mean being listened to and heard. This gave us strong guiding insights onto the importance of managing the community conversations in terms of immediacy, frequency, and quality.

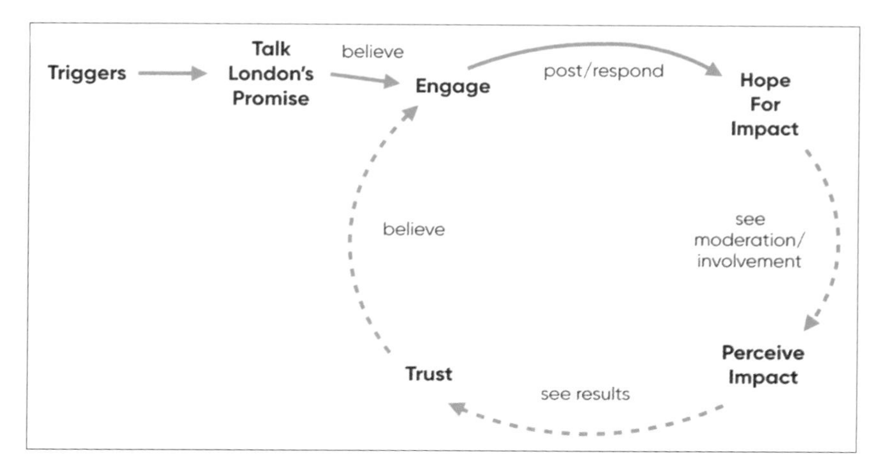

FIGURE 3.8 The cycle of trust shows the steps of getting the job done and making an impact.

CLARIFYING THE SYSTEM OF PROGRESS

By combining the JTBD Switch interview with more traditional UX research methods, we were able to get clear and detailed feedback on the new designs, including specific interactions and functionality. More importantly, we gained clarity on which elements would help to build or erode trust through the lens of the users' core job: achieving impact. This gave us very clear design direction on how we could turn a "*hope* for impact" into a *certainty* of impact, thereby building further trust and ongoing engagement.

Steph Troeth is an independent UX strategist and researcher and has been using JTBD since 2013. Previously, she spear-headed European customer research with MailChimp, led design research at the Telegraph, *and was the Head of Research at digital agency Clearleft in Brighton, UK. Currently, she is a contractor at Google UK as a senior UX researcher.*

Recap

Jobs aren't made up; they are discovered. Interviewing is a cornerstone of JTBD. Traditional research approaches, such as surveys, focus groups, and marketing studies, fail to uncover goals and needs. Instead, you should rely on firsthand qualitative interviews. Ask questions that get participants talking about their goals and their process.

An alternative technique to discovering jobs is Switch interviewing. This technique was developed by Bob Moesta and Chris Spiek, who were focused on re-creating a timeline around a recent purchase. But Switch interviews are not about product preference or satisfaction: they are focused on buying decisions and underlying motivations to seek progress by getting a job done. The technique is an example of a JTBD-as-demand-generation perspective.

After a timeline is complete, use the Four Forces technique to understand reasons for switching ways of working. There are four forces:

- **Problems** that push people away from an old solution.
- New solutions that **attract** them toward a different way of working.
- **Anxiety** that pushes people away from new solutions.
- **Habits** that pull them to stay with existing ways of working.

Mapping helps you sort out your research and provide insight into the process of carrying out a job. A job map is a chronological diagram of the main stage or subgoals of a job. Use the job map to find opportunities, compare competing solutions, and plan a long-term roadmap.

CHAPTER

4

Defining Value

IN THIS CHAPTER, YOU WILL LEARN ABOUT THESE PLAYS:

- How to find unmet needs
- How to create goal-based personas
- A new way to compare competing solutions
- How to define a value proposition

The product marketing manager at a company I once worked for stood up in a meeting to present his strategy. He proceeded to describe the top customer needs that we should support with our solutions. I was thrilled to see him align to a customer-centric model.

But something wasn't quite right. Customers don't talk about using a "system of record," as he was describing. I also thought the categories in his model seemed unnatural—our customers didn't necessarily make the same distinctions he posited, or so I thought.

So I asked, "Where did the model come from?" The product marketing manager responded casually, "Oh, we brainstormed it yesterday."

It turns out that he and his team had devised a theory of customer needs off the tops of their heads. Sure, they had spoken to customers in the past and had a general awareness of what happened in the real world. But they had devised a confected model of behavior and then defined their strategy based on it.

You should avoid this type of strategic improvisation. Instead, strive to connect the outside world with internal decision-making based on systematic models. Once you've conducted primary research, then consider how you can define the value you're going to create with proven methods. The resulting models are not only strategic in nature, but they also serve to form a common perspective and starting point for other activities downstream in the value-creation cycle.

PLAY ➤ Find Underserved Needs

Solutions that target unmet needs have a higher chance of adoption and market success, accordingly. But any given main job may have dozens or even a hundred or more needs. Prioritizing them is key. In principle, the approach is straightforward: find needs that are important and not currently well satisfied.

Figure 4.1 shows the basic idea of unmet needs with a simple matrix. The horizontal axis shows how job performers rate each need statement for satisfaction, from low to high. The vertical axis shows how important each need is, from low to high.

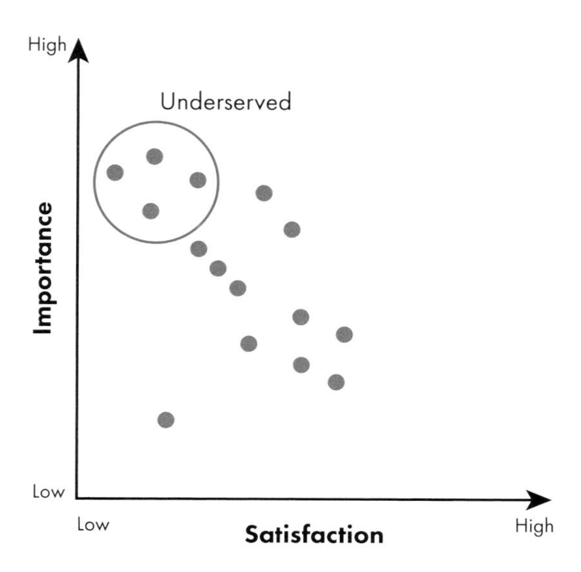

FIGURE 4.1 Customer needs that are important and not satisfied present the biggest opportunity.

Outcome-Driven Innovation

Tony Ulwick has done significant work developing an approach to quantify and pinpoint underserved needs. His company, Strategyn, pioneered a method called *Outcome-Driven Innovation* (ODI), which offers a complete process for finding strategic opportunities. They believe that if you know how the customer measures value, you can provide successful solutions in a measured, controlled way.

In ODI, "needs" are synonymous with "desired outcomes." The formulation of desired outcome statements is specific and precise.[1] There are four elements in each statement:

- **Direction of change:** How does the job performer want to improve conditions? Each desired outcome statement starts with

1. For more details, see Anthony Ulwick and Lance Bettencourt, "Giving Customers a Fair Hearing," *MIT Sloan Management Review*, April 2008. Note that Ulwick has refined his approach since publishing this article. For instance, within the ODI method, desired outcome statements now only show a downward or minimizing direction to provide more accurate responses.

a verb showing the desired change of improvement. Words like "minimize," "decrease," or "lower" show a reduction of the unit of measure, while words like "maximize," "increase," and "raise" show an upward change. (Note that "minimize" has been found to be more precise language since people can imagine what zero looks like. Maximize is fuzzier, without a bounding limit.)

- **Unit of measure:** What is the metric for success? The next element in the statement shows the unit of measure the individual wants to increase or decrease. Time, effort, skill, and likelihood are a few typical examples. Note that the measure may be subjective and relative, but it should be as concrete as possible.
- **Object of the need:** What is the need about? Indicate the object of control that will be affected by doing a job.
- **Clarifier:** What else is necessary to understand the need? Include contextual clues to clarify and provide descriptions of the circumstance in which the job takes place.

Here are some examples:

- Minimize the time it takes to summarize conference insights, e.g., as notes, presentations, reports, etc.
- Reduce the time it takes to get ingredients ready when preparing a meal.
- Minimize the time it takes to gather documents when preparing taxes, e.g., pay stubs, expenses, receipts, etc.

The normalization of your desired outcomes allows you to quantify their relative importance. After extensive qualitative research to derive the desired outcomes, you can launch a survey to pinpoint your opportunity from the job performer's perspective. Note that finding unmet needs using ODI is a precise process that requires rigorous investigation. I encourage you to learn more about the method as practiced by Strategyn by following the resources cited in this book. Next, you'll find a simplified overview of the basic approach:

STEP 1 ➤ Gather all desired outcomes.

From your qualitative research with job performers, identify a complete set of needs. You'll know when you've uncovered all needs when you start hearing the same ones over and over. This may take a dozen or two interviews, or more in complex situations. In order for the process to be effective, it needs to be comprehensive and as granular as possible.

STEP 2 ➤ Formulate desired outcome statements.

Check the form of each statement to ensure consistency. Have others in your team review and refine the statements to accurately reflect what job performers expressed in their interviews. Remove redundant statements and fill in the missing gaps.

STEP 3 ➤ Survey job performers.

Create a survey that consists of desired outcome statements as the main items. Pair each statement with two scales: one for importance and one for satisfaction. Figure 4.2 illustrates the basic arrangement for each questionnaire item based on the example job, *file taxes*.

1. Minimize the time it takes to gather documents

	Very low									Very high
	1	2	3	4	5	6	7	8	9	10
A. How important is this to you?										
B. How satisfied are you with ability to get this done?										

2. Maximize the likelihood of getting a return

	Very low									Very high
	1	2	3	4	5	6	7	8	9	10
A. How important is this to you?										
B. How well is this currently being satisfied?										

FIGURE 4.2 The survey for quantifying underserved needs rates each statement for importance and for satisfaction.

Administer the survey to job performers. The broader the domain of your main job, the more participants you'll need. The low end of your sample size should be more than 150, extending into the thousands from there. But a good rule of thumb is to have at least twice the respondents as the desired outcome statements.

STEP 4 ➤ Find opportunities.

Following ODI, calculate your opportunity. First, find the satisfaction gap, or importance minus satisfaction. Then add the satisfaction gap to the overall importance score, as shown in Figure 4.3. The resulting number will fall on a range of zero to 20. The higher the number, the bigger the opportunity will be.

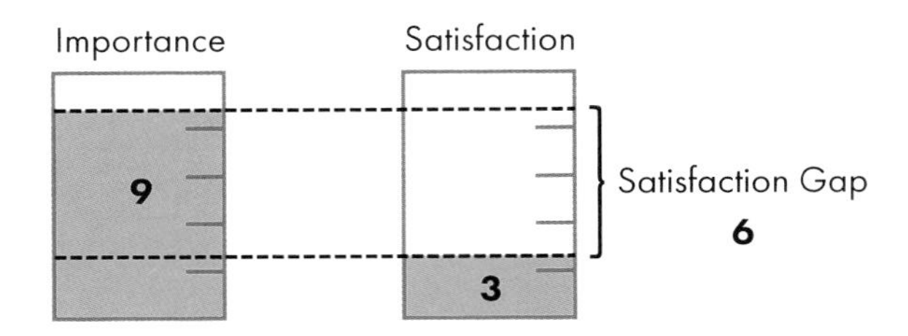

Importance + Satisfaction Gap = **Opportunity Score**
9 + 6 = **15**

FIGURE 4.3 ODI pinpoints underserved needs by calculating the opportunity score.

You may also find it helpful to plot the scores of the need statements on a graph like the one shown in Figure 4.3. This will show you a distribution of rankings visually.

While finding underserved needs following ODI might seem straightforward in principle, it's difficult to execute in practice. First, ODI requires that *all* desired outcomes in getting a job done must be collected. Second, the survey requires a large sample size in order to get valid results. Finally, it's challenging to get respondents to answer all of the questions with their full attention, and they must be incentivized to do so, making the process costly.

That being said, cutting corners is also problematic. If you don't have a complete set of consistently formulated needs and don't survey the right sample of people (namely, actual job performers), the results will be unreliable or, worse, misleading.

An alternative to prioritizing needs is to prioritize steps of the job process. Take the steps from your job map and survey them on the steps, rating each for both importance and satisfaction. Then plot each step on a simple 2 × 2 matrix based on the scores each received. From this, you can see underserved job steps instead of underserved needs. The matrix provides insight into what part of the process has the most strategic opportunity.

Related Approaches

Dan Olsen, consultant and author of *The Lean Product Playbook*, independently developed a very similar approach using an importance versus satisfaction matrix.[2] For example, let's say you were trying to find opportunities for Uber, the car ride service. Olsen recommended that you first find all of the service demands that people might have, e.g., clean cars, timely pickups, etc.

Similar to ODI, the next step would be to survey people who had used Uber by importance and satisfaction. But instead of a scale of 1–10 for both, Olsen suggested using a different scoring system. Importance should be placed on a scale of 1–5 and satisfaction on a scale of 1–7. Importance could best be seen as a polar scale. That is, the concept of negative importance didn't really make sense. The scale of 1–5 went from low to high. Negative satisfaction, however, was possible. So, for satisfaction, Olsen used a Likert scale with the middle value equaling a neutral state.

Olsen placed the resulting scores in a matrix, as seen in Figure 4.4.

2. Dan Olsen, *The Lean Product Playbook* (Hoboken: Wiley, 2015).

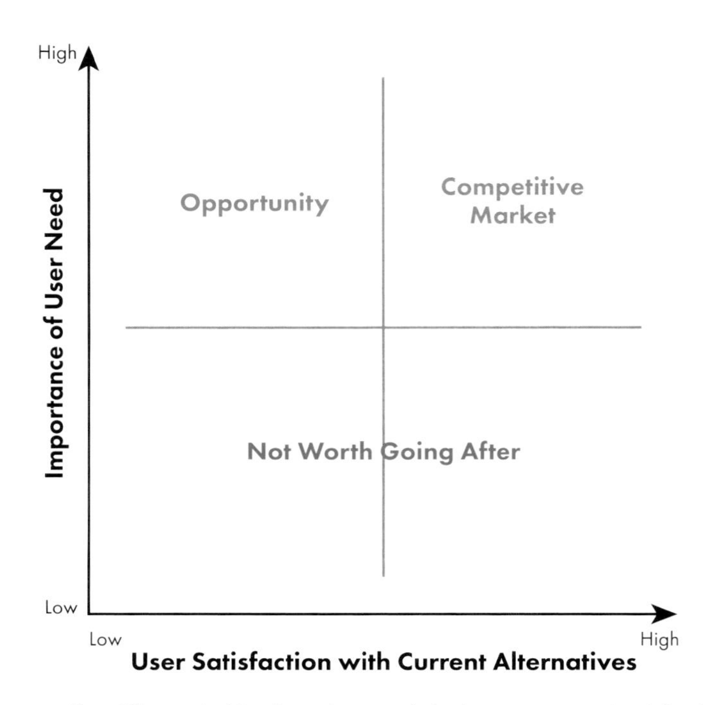

FIGURE 4.4 Dan Olsen prioritized product needs by importance and satisfaction.

Starting with an existing solution in mind, Olsen's technique lies within the solution space of JTBD. Importance and satisfaction were based on features or service aspects. From this perspective, the approach was closer to the voice of customer research, which considered feedback from customers about a given product or service, rather than JTBD. Still, there was a common point of attempting to find opportunity by ranking importance and satisfaction for each need.

Interestingly, Olsen suggests that the importance versus satisfaction matrix can be used with a sample size of zero. In other words, after doing research to gather the needs from job performers, a team can simply make assumptions about the relative position of each need on the graph. From that analysis, possible solutions can be conceived and then tested in experiments, tying JTBD to Lean methods. Overall, market feedback comes from reacting to a prototype rather than to a desired outcome statement.

For instance, imagine you worked for a startup that intended to help people *take medication at a specific time of the day.* After qualitative research uncovered the need, you could take those needs and prioritize them on the importance-satisfaction matrix relative to each other making assumptions. Then you could form a hypothesis you believed to have the most opportunity and conduct experiments to test that hypothesis with solutions. For example, you might want to see if electronic reminders on a mobile device would address the need.

In another related approach outlined in the book *The Innovator's Guide to Growth*, Scott Anthony and his coauthors showed that marketplace opportunity comes from understanding the customer job:[3] "To identify opportunities to create new growth, look first for important "jobs" that people can't get done satisfactorily with current solutions."

This doesn't just apply to creating new offerings. JTBD can help revitalize growth in existing businesses. The authors wrote: "Jobs-based thinking can restart growth by helping companies shake up commoditized markets and highlighting opportunities to revive even the most moribund of products."

One specific method to evaluate opportunities from a jobs-based view of the market is the jobs scoring sheet. After doing primary research to uncover jobs, including interviewing and observations, Anthony and his colleagues recommended prioritizing jobs along several dimensions. For each job, ask these three questions and rate each on a scale of 1–5:

- Is the job important to the customer? (1 = not important, 5 = very important)
- Does the job occur relatively frequently? (1 = infrequently, 5 = very frequently)
- Is the customer frustrated by the inability to get the job done with today's solutions? (1 = frustrated, 5 = not frustrated)

3. Scott Anthony, Mark Johnson, Joseph Sinfield, and Elizabeth Altman, *The Innovator's Guide to Growth* (Boston: Harvard Business Press, 2008).

Then create a score for each job statement using the following equation: (importance) + (frequency) × (frustration). Finally, rank each job statement by its score. The higher the score, the greater the opportunity.

Table 4.1 reflects a simple worksheet for following their approach. It's best filled out based on survey data from job performers, but can also be used as a worksheet for teams to reflect on opportunities based on their existing knowledge and understanding of customer jobs.

TABLE 4.1 THE JOBS SCORING SHEET

JTBD (OUTCOME OR STEP)	IMPORTANCE	FREQUENCY	FRUSTRATION	SCORE	RANK
1.					
2.					
3.					
4.					
5.					
6.					
7.					

Finally, there are even more informal ways to use the type of thinking outlined in this play. For instance, I have used informal ways to work with unmet needs based on the importance-satisfaction matrix. During one workshop I conducted, the group was required to first generate need statements based on previous research. Then we prioritized them only by assumed importance first. From the cluster of needs that were deemed to be more important, I instructed each group to decide which they felt were most unsatisfied. Teams then selected a need and ideated solutions around it.

LEARN MORE ABOUT THIS PLAY: **FIND UNDERSERVED NEEDS**

Anthony Ulwick, "Turn Customer Input into Innovation," *Harvard Business Review* (January, 2002).

> This is Ulwick's landmark article that outlines a complete process for finding unmet needs. He provides the rationale behind its approach and detailed steps to follow. Ulwick's approach is hard to replicate, but represents precise problem-space needs understanding.

Dan Olsen, *The Lean Product Playbook* (Hoboken: Wiley, 2015).

> Chapter 4 of Olsen's book is dedicated to techniques for identifying underserved customer needs. He describes the importance-satisfaction framework in detail and provides many examples. Olsen does not present a JTBD method or technique specifically, but his thinking and approach overlap with similar thinking.

Scott Anthony et al., *The Innovator's Guide to Growth* (Boston: Harvard Business Review Press, 2008).

> This book contains a wealth of strategic and practical information for creating successful innovation programs. The authors bring collectively decades of experience to the table. JTBD plays a significant role in their approach, particularly during up-front problem definition and deciding on strategic targets.

WHICH COMES FIRST: NEEDS OR TECHNOLOGY?

In his 2009 article "Technology First, Needs Last," Don Norman provoked the design community with a polemic assertion:[4]

> I've come to a disconcerting conclusion: design research is great when it comes to improving existing product categories but is essentially useless when it comes to new, innovative breakthroughs... New conceptual breakthroughs are invariably driven by the development of new technologies.

A pioneer in human-centered design, Norman seemed to defy his own school of thought with this contention. But, he has a point: when it comes to revolutionary technologies—history shows that technology often came first.

In that same year, Sarah Miller Caldicott, grandniece of Thomas Edison, wrote extensively about her great uncle's work. After studying his approach to innovation for years and poring through his notes and papers, she concluded:[5]

> Edison realized that by understanding customer needs first, he could invent useful products more efficiently than he could otherwise. Edison's trained teams visited people in their homes and watched how they used their current lighting products.

4. Don Norman, "Technology First, Needs Last," jnd.org December 5, 2009.
5. Sarah Miller Caldicott, "Ideas-First or Needs-First: What Would Edison Say?" (white paper, Strategyn, 2009).

This begs the question, which comes first: technology or needs?

Everett Rogers, originator of innovation adoption theory, offers some insight here. In his landmark book *Diffusion of Innovations*, Rogers points out:[6] "An individual may develop a need when he or she learns that an innovation exists. Therefore, innovations *can* lead to needs, as well as vice versa."

Like Rogers, I believe the answer to the question is both. Said another way, it doesn't matter where impetus for innovation comes from: it's an iteration between technology and needs. If inspiration comes from a new discovery or invention, that can lead to innovation. On the other hand, human needs can point to opportunities, too.

However, always keep the end point of innovation in mind: the end always lies with the people who adopt the innovation. It's up to them to decide whether they want to adopt an innovation or not.

Ultimately, it's not a question of where the inspiration for innovation originates, but rather where it ends: with human desires and wants. JTBD seeks to reduce the chance of non-adoption by focusing on solutions that people actually need. Jobs thinking strives to identify what outcomes are sought and connect those to discovery and invention. What's more, JTBD gives us a consistent language and approach to align teams around customer needs, which, in turn, raises your chances of adoption.

6. Everett Rogers, *Diffusion of Innovations*, 5th ed. (New York: Simon & Schuster, 2003).

PLAY ➤ Create Goal-Based Personas

Personas are archetypal representations of users. They are used in various areas of a business to instill customer-centric decision-making—from marketing to sales to design. A persona is essentially a communication tool—a way to summarize information about customers in a way that is accessible to everyone on a team.

Some practitioners within the JTBD community have called to abandon personas. They point out the flaws in personas that are based on demographic, psychographic, behavioral, and/or attitudinal schemes. But while these aspects may be central to traditional marketing approaches, they don't inform innovation and design.

It's a misconception, however, that JTBD replaces the need for personas entirely. In fact, given the multiple actors in any given JTBD ecosystem, it stands to reason that personas can represent the various roles. What's more, not all job performers are the same, and personas can be used to illustrate different types of performers.

To help understand how to create personas based on people's goals, the best place to start is the work of Alan Cooper. His approach, introduced in his book *The Inmates Are Running the Asylum,*[7] bases personas not on demographics but rather their intended outcomes. "Personas are defined by their goals," he writes.

In his follow-up book, *About Face 2.0,*[8] Cooper presents complete instructions for creating personas as part of his Goal-Directed Design (GDD) method. Although GDD and JTBD were developed independent of one another, there is overlap. Both focus on the goals that people

7. Alan Cooper, *The Inmates Are Running the Asylum* (Indianapolis: SAMS, 1999).
8. Alan Cooper and Robert Reimann, *About Face 2.0: The Essentials of Interaction Design* (Indianapolis: Wiley, 2003).

have. However, GDD is a design method specifically created to design software user interfaces. As a result, the approach refers to users of a predetermined system and operates in the solution space.

Nonetheless, Cooper's approach serves as a good basis for creating goal-based personas, summarized in the steps below:

STEP 1 ➤ Interview users.

Personas are based on qualitative interviewing, similar to the interview technique outlined in Chapter 3, "Discovering Value." In fact, you can interview for both jobs and personas in the same session. Rather than asking users about their preferences or desires, focus on their intent, as well as what frustrates them and what success looks like.

A dozen hour-long interviews usually suffice. In more complex situations, it may be necessary to interview two or more dozen people. The aim is to get enough feedback from enough people to find clear patterns. You'll know when you've done enough interviews when you can start predicting how people will respond.

STEP 2 ➤ Map interviews to variables.

Following Cooper's approach, the next step is to map interviews to so-called behavior variables. Think of variables as needs with two end points that create a range. Variables correspond approximately to the circumstances in JTBD, but can include other factors as well. For online shopping, for example, the variables might be frequency of shopping, degree of enjoyment, and price vs. service orientation.

Place each interview on the range you create relative to each other. Figure 4.5 shows how a set of five interviews might be mapped to just two variables uncovered during research.

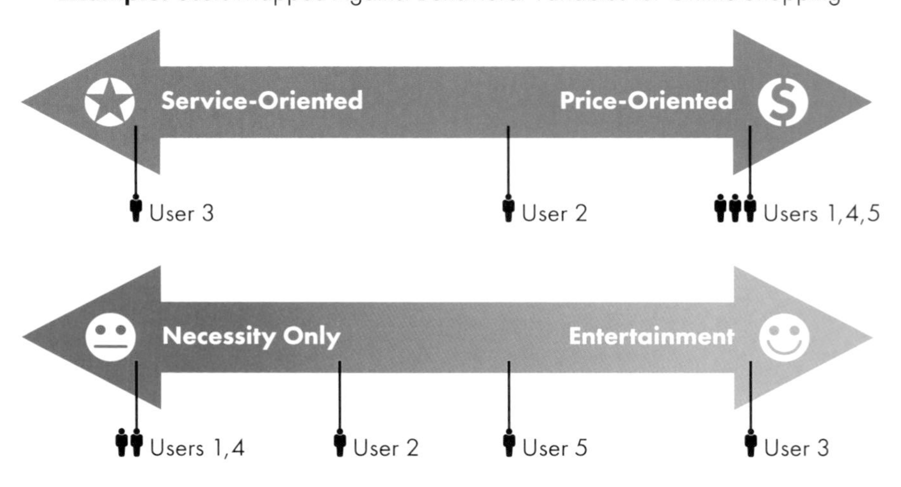

FIGURE 4.5 After conducting research, map behavior variables of all participants.

STEP 3 ➤ Identify patterns in goals.

After mapping interview participants to the variables, look for clusters of similar behavior. A group of participants that cluster similarly across six to eight variables represent a significant behavior pattern.

This isn't a statistical exercise, but rather the search patterns in causation between the variables. In the example shown in Figure 4.5, for instance, participants 1 and 4 both shop for necessity and are price conscious. It stands to reason that these goals influence each other.

STEP 4 ➤ Describe the resulting persona.

For each cluster, create a separate persona similar to the example in Figure 4.6. The basic description should include their most important goals and their relevant circumstances. You should reflect a typical workday or circumstances when reaching the goal, as well as work-arounds and frustrations. Avoid adding irrelevant detail: just one or two personal aspects can make a persona feel real. Focus on the patterns of behavior instead. In the end, the personas can be used to prioritize or focus efforts to create an appropriate solution.

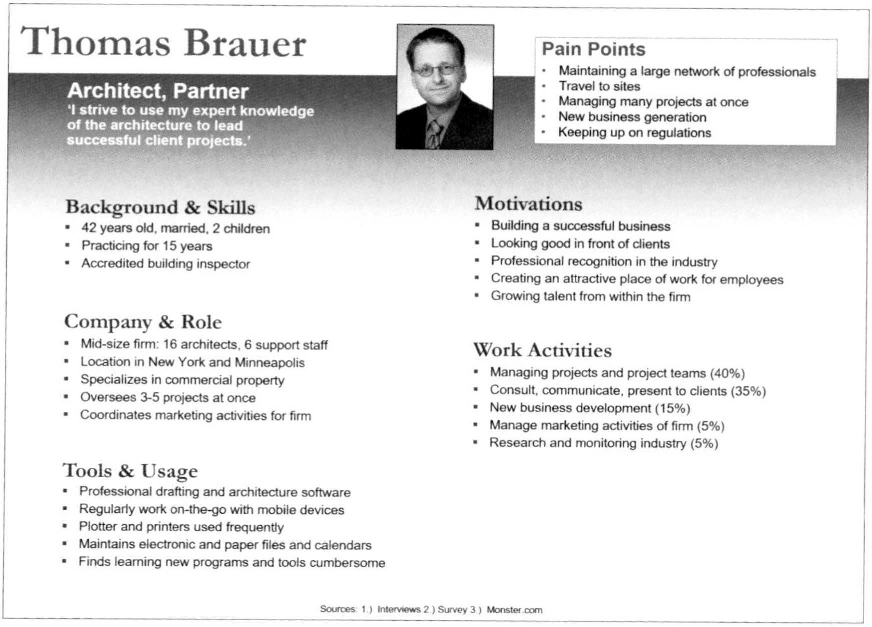

Thomas Brauer

Architect, Partner
'I strive to use my expert knowledge of the architecture to lead successful client projects.'

Pain Points
- Maintaining a large network of professionals
- Travel to sites
- Managing many projects at once
- New business generation
- Keeping up on regulations

Background & Skills
- 42 years old, married, 2 children
- Practicing for 15 years
- Accredited building inspector

Company & Role
- Mid-size firm: 16 architects, 6 support staff
- Location in New York and Minneapolis
- Specializes in commercial property
- Oversees 3-5 projects at once
- Coordinates marketing activities for firm

Tools & Usage
- Professional drafting and architecture software
- Regularly work on-the-go with mobile devices
- Plotter and printers used frequently
- Maintains electronic and paper files and calendars
- Finds learning new programs and tools cumbersome

Motivations
- Building a successful business
- Looking good in front of clients
- Professional recognition in the industry
- Creating an attractive place of work for employees
- Growing talent from within the firm

Work Activities
- Managing projects and project teams (40%)
- Consult, communicate, present to clients (35%)
- New business development (15%)
- Manage marketing activities of firm (5%)
- Research and monitoring industry (5%)

Sources: 1.) Interviews 2.) Survey 3.) Monster.com

FIGURE 4.6 Goal-based personas include minimal demographic information and focus instead on goals and frustrations.

Overall, Cooper's approach provides a standard for creating goal-based personas. However, because the method looks at users of a predetermined system, it also falls within the solution space. What's more, as a specific software interface design method, GDD has had little influence outside of design communities.

RELATED APPROACHES

Instead of differentiating based on behavioral variables, use the circumstances you uncover during your research. Frame each as a range and map interviews to them. Then find different clusters of job performers. The process is similar to the one outlined previously.

- **Interview job performers.** Instead of interviewing users, interview job performers. As you uncover their jobs and needs, you can also determine the different circumstances that matter most.

- **Map interviews to circumstances.** Map the interviews to the circumstances. Rather than a range, as in GDD, simply take inventory of which interviews correspond to each circumstance.
- **Find patterns.** Look for logical clusters of circumstances that show causality.
- **Describe the persona.** Use a similar format to the persona shown in the previous example.

In addition to circumstances, Stephen Wunker, Jessica Wattman, and David Farber recommend including two other variables they call drivers in their book *Jobs to Be Done.*[9] They recommend considering *attitudes*, or the personality traits relevant to getting a job done, as well as people's *backgrounds*, or long-term context that affects decision-making. Together, these drivers—circumstances, attitudes, and backgrounds—form the basis of segmenting individuals into different segments based on their goals.

It's also possible to create personas for roles other than the job performer. Chief among these is a buyer persona. You might also find that personas are needed for approvers, technicians, or people who are beneficiaries of job outcomes.

To speed up the process, try creating proto-personas, a quick format for creating personas that can be used in a group exercise. Also referred to as *assumption personas*, these are depictions of job performers based on what you know today.

Proto-personas can be created on a simple grid with four squares, as shown in Figure 4.7. There are four main elements:

- **Name and sketch:** Pick a name that is memorable and include a sketch.
- **Demographic details:** Include factors that are relevant to the main job.

9. Stephen Wunker, Jessica Wattman, and David Farber, *Jobs to Be Done: A Roadmap for Customer-Centered Innovation* (New York: AMACON, 2016).

- **Behaviors:** Indicate the key actions the person takes when performing the job.
- **Needs:** Indicate what the person needs and what difficulties he or she might encounter.

Name and Sketch	Behaviors and Actions
Mary	Reads newspaper daily Subscribes to weekly news magazine Watches news on TV 1–2 times a week Gets digest news as email (e.g., Daily Beast)
Demographic and Psychographic Details	**Needs and Pain Points**
College grad 32 years old, single Full-time marketing manager	Needs short, consumable news stories Difficulty reading news on mobile device Overwhelmed by sources of news Needs to be able to trust a source for accuracy

FIGURE 4.7 Use the proto-persona format in a collaborative group session.

Overall, the success of using personas can vary. The key is not only to base them on goals, but also to make them visible and interactive. Involve your team in their creation for greater buy-in. Otherwise, it's not uncommon for personas to get shelved and have limited utility across time. Don't fall into this trap, and make your personas relevant and actionable.

LEARN MORE ABOUT THIS PLAY:
CREATE GOAL-BASED PERSONAS

Alan Cooper and Robert Reimann, *About Face 2.0: The Essentials of Interaction Design* (Indianapolis: Wiley, 2003).

> This thick volume is specifically geared to UI designers and includes a wealth of information on interaction design. The first chapter outlines Cooper's approach to creating goal-based personas in detail. The language used throughout assumes a target outcome of software products, limiting its apparent relevance to other disciplines and other types of products and services.

Kim Goodwin, *Designing for the Digital Age: How to Create Human-Centered Products and Services* (Indianapolis: Wiley, 2009).

> Goodwin is a clear thought leader in UX design, and her work in personas deepens Goal-Directed Design as a standard. This massive volume details a complete approach to digital product design from end to end, but with significant sections on user research, modeling, and personas. She also has a lot of material on personas on the web in the form of articles and blog posts that are more accessible and approachable.

John Pruitt and Tamara Adlin, *The Persona Lifecycle: Keeping People in Mind Throughout Product Design* (San Francisco: Morgan Kaufmann, 2006).

> This is a comprehensive manual on creating personas in general. The authors don't necessarily follow a JTBD approach for personas but do offer discussions of needs and goals as a basis for creating personas.

PLAY ➤ Compare Competing Solutions

Traditional competitive analysis almost always involves a technical comparison of product specifications and features, yet the analysis is conducted without knowing how customers measure value or how much value competing features deliver to the customer. This is the problem *and* the myth that misleads: companies are *not* competing against other companies or their products. They are competing *for* the customers, and their one goal is to create value for them. And there is only one way to do that: by offering a product or service that is better than any other at helping them get their jobs done.

Part of the problem, too, is a myopic view of the competition. Just consider this quote from Scott Cook, the founder of Intuit:[10]

> The greatest competitor [to tax software] we saw was not in the industry. It was the pencil. The pencil is a tough and resilient substitute. Yet the entire industry had overlooked it.

Think about it: tax software doesn't just compete with other tax software solutions. It competes with everything a customer uses to get a job done. When preparing taxes, there are many subcalculations and figures you have to bring together. Grabbing for a pencil is often the easiest solution. So from a JTBD perspective, tax software competes with pencils.

Since JTBD is solution agnostic, it enables a new way of viewing competition. Rather than considering providers in a defined industry only, it's possible to compare across product categories.

Companies typically conduct competitive analysis to make sure their products and services are better than those offered by competitors.

10. Quoted in Scott Berkun's book *The Myths of Innovation* (Sebastopol, CA: O'Reilly, 2007).

But reacting to feature-to-feature comparisons is no guarantee of success. Competitive analysis, when seen through a jobs-to-be-done lens, is not about head-to-head comparisons. Instead, it's about assessing how much better or worse a product is at helping the customer get a job done.

For instance, in his famous milkshake story, Clayton Christensen discusses different forms of breakfast that people might get while commuting to work. He shows that milkshakes perform better where others fail, such as a bagel, banana, and Snickers.

Typically, milkshake providers would compare their products to other milkshake providers. They'd look at the milkshake itself and make theirs thicker or thinner to differentiate it. But thickness doesn't apply to bananas, Snickers bars, and bagels.

With JTBD, you can compare products and services to user needs. This gives a different view of the competition, and it steps outside of defined product categories. Looking at just a small sample of possible needs for getting breakfast on the go, it's easy to see how well each one performs, as shown in Table 4.2.

TABLE 4.2 COMPARING SOLUTIONS BY NEED

	MILKSHAKE	BANANA	SNICKERS	BAGEL
Maximize satiety	●	○	◐	●
Increase the ease of eating	●	○	●	○
Minimize the risk of causing an accident while on the go	●	◐	●	○
Decrease the mess caused while consuming	●	◐	●	○
Increase ability to consume slowly over time	●	○	○	◐
Avoid feeling guilty about food consumed	◐	●	○	●
(List additional needs here...)	...	...	...	...

While this approach may seem straightforward on the surface of things, it's more difficult to complete in reality. First, deciding what needs to compare is tricky. Ideally, you will compare all of the needs you've uncovered. Alternatively, focusing on underserved needs only will help narrow the list. Second, in many cases there may be an endless number of solutions to compare. Deciding which to include requires a strategic selection.

Take a step back from your solution and product category, and compare the competition as your market sees it: by getting a job done. Here's how it works:

STEP 1 ➤ Determine alternatives to compare.
Start with your main job and job map. Then list all of the means by which people get the job done at each stage in the process. From this list, select the solutions that are most relevant to compare and place them across the top of the row of the table as column headers. Note that you can do this before you even have a product or service of your own on the market.

STEP 2 ➤ Determine the needs to compare.
It's possible to use the full set of outcomes to do a competitor comparison, but typically you'll focus on a subset, such as the top unmet needs. List these on the left column of the table.

STEP 3 ➤ Rank how well each solution meets those needs.
Ideally, you'd survey job performers about each of the needs you selected to compare. Set up a questionnaire similar to the one described previously for targeting underserved needs: for each need statement, provide scales for importance and for satisfaction. Then calculate the opportunity score for each and enter that into the table.

If surveying job performers isn't possible, estimate how well each one meets the needs in a team exercise. Rank the needs with "low, medium, high," and negotiate a response for each together. Just be aware that without feedback from job performers, your rankings are speculative.

STEP 4 ➤ Find your sweet spot in the competitive landscape. Determine the needs that are missed by others or where you perform better. The aim isn't to find a feature that hasn't been built, but rather what needs are being underserved. When designing a new solution or when improving an existing one, use this insight into which opportunities to tackle first.

Related Approaches

Instead of comparing need statements, compare steps in the job process. The process is identical to the comparison outlined previously, using the steps in the job process as the basis for comparison.

Table 4.3 illustrates what a comparison might look like with a subset of all steps for the job—*prepare a meal*. It compares three services: manually preparing a meal from scratch, using pre-cut ingredients (e.g., chopped vegetables), and full meal kit services with portioned ingredients. The relative level of satisfying each step is indicated with a simple low, medium, or high.

TABLE 4.3 COMPARING JOB STEPS ACROSS SOLUTIONS

	MANUAL PREPARATION FROM SCRATCH	PRE-PREPARED INGREDIENTS (E.G., CHOPPED, DICED, ETC)	FULL MEAL KIT SERVICES
Decide what to make	Low	Medium	High
Gather ingredients	Low	Medium	High
Prepare ingredients	Low	High	High
Make the meal	Medium	Medium	High
Serve the meal	Low	Low	Low
Clean up	Low	Low	Low
Store leftovers	Low	Low	Low

From just this simple comparison, it's clear that both pre-prepared ingredients and meal kit services help customers in the earlier steps of getting the job done—more so than preparing a meal from scratch.

It's also possible to graph your comparison to show advantages and disadvantages clearly. For instance, I previously worked for a provider of online content. At the time, we were trying to understand why people preferred print resources over digital resources. From interviewing dozens of people, we found a key set of needs that reflected the difference.

On a graph, we showed how the needs compared in terms of being met or not met by each solution. Figure 4.8 shows a modified version of that graph. We then hypothesized about what would need to be true for people to use our online content solutions more, labeled "New Online Experience" in the diagram. It became clear that we needed to cover more of the needs that people had reading and annotating documents, as well as comparing sources and navigating them, seen on the right side of Figure 4.8.

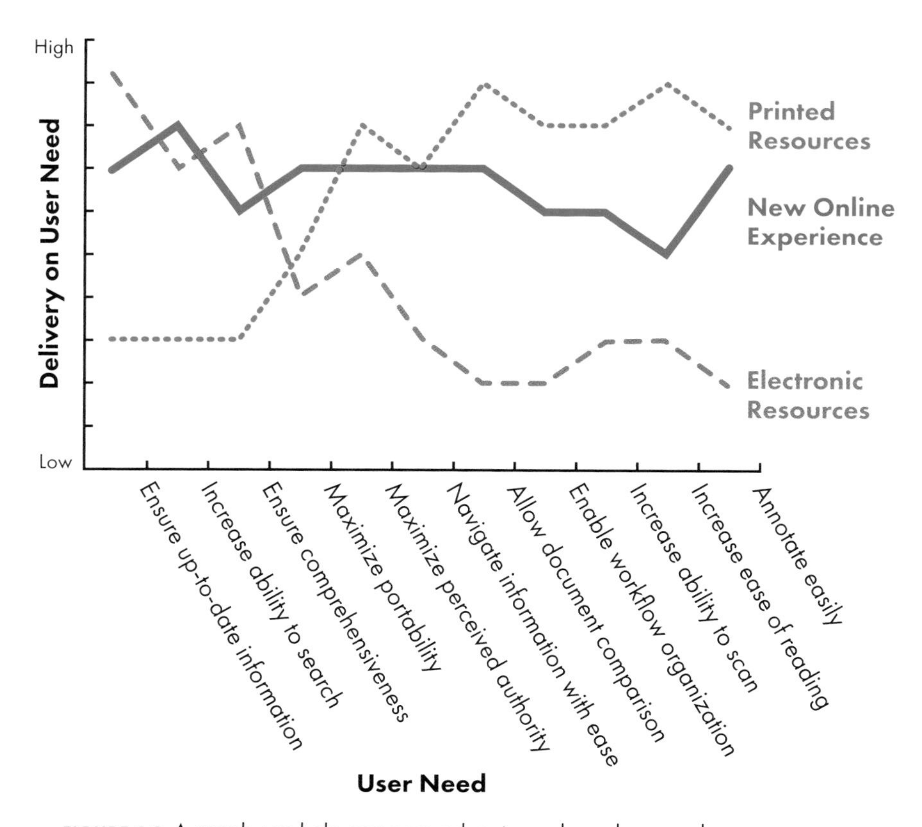

FIGURE 4.8 A graph can help compare advantages based on needs.

LEARN MORE ABOUT THIS PLAY:
COMPARE COMPETING SOLUTIONS

Stephen Wunker, Jessica Wattman, and David Farber, "Competition," Chap. 7 in *Jobs to Be Done: A Roadmap for Customer-Centered Innovation* (New York: AMACON, 2016).

> This chapter outlines the advantages of viewing competition through the lens of JTBD. Companies don't challenge the established views of their industries often enough, the authors argue. They point to three factors to consider during competitive assessments: advantages delivering against the job to be done, flexibility to adjust to meet customer needs, and rivals' impact on marketplace perceptions.

Des Traynor, "Understanding Your Real Competitors," Chap. 2 in *Intercom on Jobs-to-Be-Done* (Self-published, 2016).

> In this short essay, Traynor lays out key points in reframing how to look at the competitor differently with JTBD. Product categories put blinders on the traditional view of competition. He details three types of competition: direct, secondary, and indirect competition. "When you're blinded by thinking your competitors only exist in the exact same tool category you're in, disruption and destruction will come from oblique angles," Traynor writes.

PLAY ➤

Define a Jobs-Based Value Proposition

A value proposition is a promise an organization makes to its customers. Although often associated with a specific statement, a value proposition is much more than a marketing device. It ultimately explains the benefit provided to the customer.

Value in this context is a perceived attribute. What a company thinks is very valuable may not be seen as such by consumers of its products or services. JTBD helps pinpoint those aspects that people value most, namely, meeting their needs.

Value Proposition Canvas

Alexander Osterwalder and his team at Strategyzer have developed a systematic approach for arriving at a value proposition. The Value Proposition Canvas (VPC) is designed to facilitate discussion around value propositions and align them to customer goals. Overall, the intent is to match the value you intend to create to the value that customers perceive as most beneficial. Accordingly, the VPC has two parts, reflected in the diagram in Figure 4.9.[11]

On the right is the customer profile with three components: the customer jobs, their pains, and their gains. Together, these elements represent the customer value profile.

The left half of the canvas details the features of your value proposition. There are three elements: your products and services, pain relievers, and gain creators. Together, these aspects reflect an understanding of the offering you're providing.

11. The Value Proposition Canvas is reprinted here with permission from Strategyzer. Learn more on their website: www.strategyzer.com

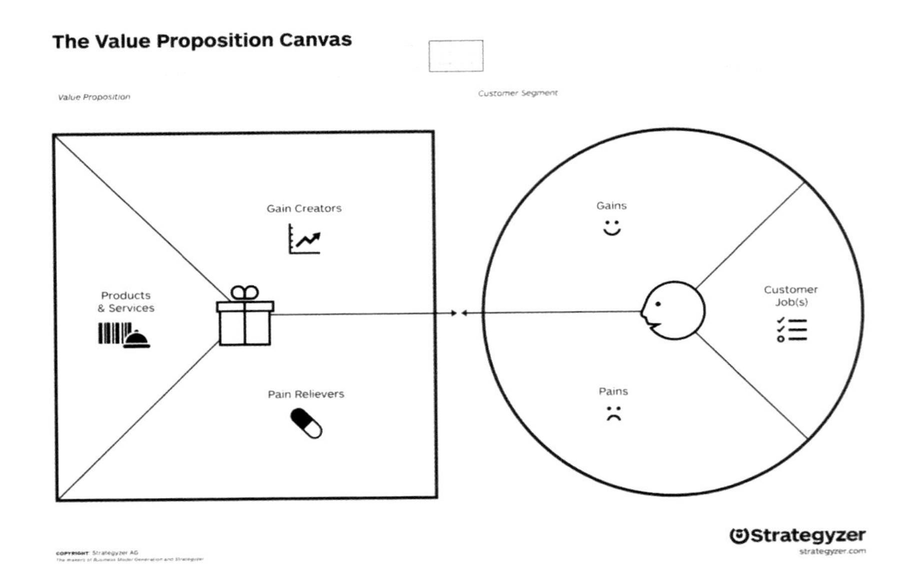

FIGURE 4.9 The Value Proposition Canvas created by Alexander Osterwalder and his company, Strategyzer.

Note that Osterwalder doesn't provide specific guidance on how to uncover and formulate jobs to be done, nor on how to prioritize them. His process is much more intuitive. Using some of the techniques in this book, however, you can come to the VPC with key jobs and needs in hand.

By mapping the left side to the right side (solution space understanding), you can make explicit how you are creating value for your customers. When the pain relievers and gain creators correlate to the pains and gains of your customers, you have a potentially strong fit. The process is collaborative and fairly straightforward.

STEP 1 ➤ Understand the customer profile.
Start by discussing your customers and the jobs they want to get done. List the main job along with crucial social and emotional aspects. In the end, you should have a limited list of jobs here, about two to five on average.

List and discuss more important pains, or the negative aspects that people have in relation to getting the job done. These are the frustrations, challenges, risks, and struggles you uncover in your research. Avoid talking about specific solutions and discuss the general pains that people have. Focus on the most severe pains, typically the top dozen or so.

Gains are not simply the opposite of pains. They are the positive benefits people want to have from completing a job, including positive emotions, surprises, and even ambitions. List the most relevant desired outcomes that people have when performing the job independent of a given solution.

Typically, you can fill out the right side of the VPC intuitively, based on your prior research. However, your team may find more interviews are necessary to get more granular on the gains and pains. Fill in any gaps in knowledge with additional investigation.

STEP 2 ➤ Discuss the solution profile.
Start with the products and services you offer or plan to offer. Then make it explicit how your products and services might alleviate the pains you identified. Consider the entire job process here—before, during, and after getting a job done. Next, consider how your products and services create customer gains and list them under Gain Creators.

STEP 3 ➤ Ensure fit between the customer and the solution.
Map the pain relievers and gain creators to the pains and gains respectively. This shows which of the pains you're addressing, and which gains you are building. Draw lines connecting them as your team discusses each point.

According to Osterwalder, you have problem-solution fit when the features of your value proposition (on the left) match the characteristics of your customer profile. In other words, mapping left to right on the VPC shows problem-solution fit.

If you've done Switch interviews, use your Forces analysis to determine the pains and gains. The factors that push customers toward a new solution translate well to gains, and the forces pulling customers back represent pains.

Looking at the other half of the canvas, you can then gauge potential product-market fit. Validate this match with the market. Often, your assumptions aren't true, and real market feedback shows weaknesses in the value you propose to create. To ensure product-market fit, conduct experiments to validate your value hypothesis.

STEP 4 ➤ **Form a value proposition statement.**

After validating the fit of your value with the market, encapsulate its essence in a single statement. There are many formats for doing this, usually similar to the pattern below:

- For (target job performers)
- Who are dissatisfied with (the current alternative)
- Our solution is a (product or service)
- That provides (key problem-solving capability)
- Unlike (the product alternative)

Overall, the VPC clarifies the value that an organization provides with JTBD as a starting point. Ultimately, people value getting their job done. Aligning to the job from the beginning helps ensure that it is the value that people really want.

LEARN MORE ABOUT THIS PLAY:

DEFINE A JOBS-BASED VALUE PROPOSITION

Alexander Osterwalder et al., *Value Proposition Design* (Hoboken: Wiley, 2014).

> This book details a practical method for defining a value proposition based on the VPC. It's written in an engaging style with many examples and techniques designed for quick reference. See also Osterwalder's landmark book *Business Model Generation* (Wiley, 2010).

Using JTBD to Align Product Strategy with Customer Needs

By Vito Loconte, UX Research Manager at Trulia

Trulia's current mission is to "build a more neighborly world by helping you discover a place you'll love to live," but that wasn't always the focus of the brand. In 2016, we were looking for our next big opportunity and a way to differentiate ourselves from the rest of the real estate category. We took a user-centered approach to figure this out and looked to our users to understand what they love about Trulia today and what their biggest unmet needs were when shopping for a home.

WHAT USERS LOVE ABOUT US

When we looked through the comments in our consumer listening posts, there was a clear reason many home shoppers kept coming back to Trulia—our Local Information. Trulia had already been providing information about the location outside of the house, such as crime statistics, commute data, and assigned schools. We saw an opportunity to capitalize on this Local Info and have that be our next big swing. The president of Trulia and the head of product were on board with this direction. Now, we'd have to figure out where exactly to focus our energy.

Like many projects before, we started by organizing a set of company-wide brainstorm sessions to come up with ideas for how we could gather and provide more Local Info to home shoppers. We came up with over 30 concepts, which were essentially solutions to location-based needs that we assumed home shoppers had. Our next step was to put these concepts in a survey and get feedback from consumers.

CONTINUES ➤

CONTINUED ➤

FROM CONCEPTS TO JOBS

Having done survey-based concept tests like this in the past, I worried that the majority of concepts would be rated positively, and we wouldn't be any closer to knowing what problems to actually focus on helping consumers solve. It was at this point that we decided to take a step back to think about all of the needs we were trying to solve with these concepts. Instead of just getting feedback on all of these concepts to understand which ones were desirable, we wanted to have a better sense of what jobs these concepts would be hired to solve.

So we went back through all of the concepts the team created and developed a set of jobs we were helping home shoppers accomplish with each of these concepts. Once we developed the set of location-related jobs, we needed to understand which of these jobs were most important and least met by the tools and services that home shoppers had access to today. We put those job statements into a survey and had home shoppers rank each of the jobs based on importance and satisfaction. We included location-related job statements, such as "discover new neighborhoods similar to those I'm already looking in"; "get a sense of what a neighborhood looks like before going there"; and "understand what it's like to live in a neighborhood from those who live there."

We used the importance-satisfaction matrix to pinpoint unmet needs, shown in Figure 4.10.

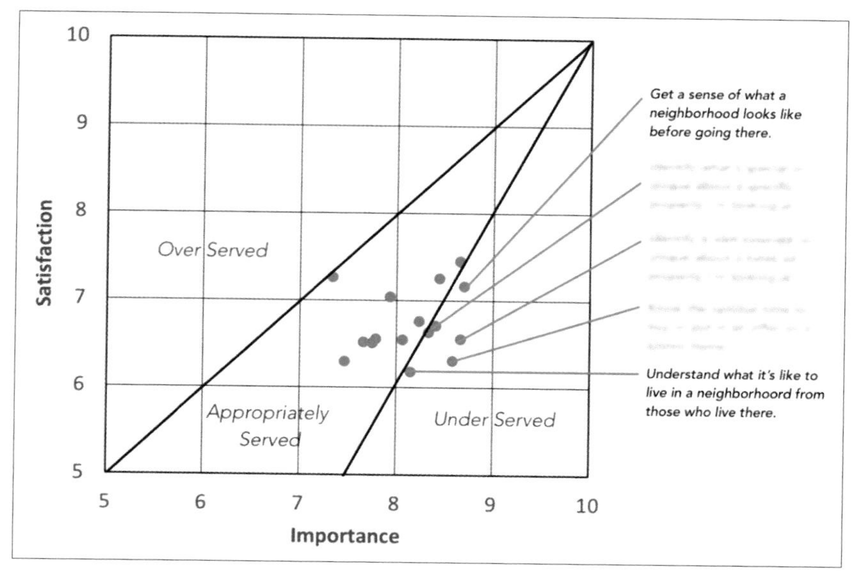

FIGURE 4.10 Trulia surveyed consumers on the importance and satisfaction of a complete set of job statements to find underserved needs.

ALIGNMENT

Based on this survey, we decided our first focus would be "to understand what it's like to live in a neighborhood from those who live there," and we developed a strategy to start collecting user-generated content (UGC) from our current user base as our first step. By focusing on this job, we knew that we were satisfying an unmet consumer need and, at the same time, we weren't limiting ourselves to a single concept or solution. What's more, we knew that unique content like user-generated content was beneficial for SEO ranking, so focusing on this need ultimately satisfied a user need and business need at the same time.

CONTINUES ➤

CONTINUED ➤

LESSONS LEARNED

One of the biggest lessons I learned from this experience was that you don't need to have the whole company on board with jobs to be done in order to start using the framework. We were able to successfully align on a product direction that was rooted in a consumer need, which was also beneficial to the business as a result of shifting our focus to the job.

Overall, the project was a huge success. We started collecting neighborhood facts and reviews from our users in 2017 and today we gather over 100,000 responses a day from people sharing what it's like to live in their neighborhood. Since the launch, we've even shifted our focus to another top-rated job from our survey—get a sense of what a neighborhood looks like before going there. We recently developed neighborhood pages that use custom photography and drone footage to give people a sense of what many neighborhoods look like without having to go see them in person.

Vito Loconte is a user researcher located in San Francisco with an M.S. in Human Factors and Ergonomics. He is an expert in various methods, including usability testing, ethnography, and JTBD research. For more on Trulia Neighborhoods, see: Mark Wilson, "Trulia Is Building the Netflix for Neighborhoods," Fast Company (2018), https://www.fastcompany.com/90214922/trulias-new-feature-is-like-netflix-for-neighborhoods

Recap

JTBD provides a framework and common language for translating insights from job performers into a model for action. The process begins with field research, but then moves into more precise modeling, both qualitative and quantitative. Before creating solutions, define the value you're going after. There are several ways you can do that.

A first step you can take is to find underserved needs. These are needs that are important to job performers but satisfying, given any and all existing solutions available to them. The importance-satisfaction matrix helps you do this.

It's also possible to define personas of actors within your JTBD ecosystem. First, segment different types of job performers based on affinity of need statements. Also, consider other roles you can define as personas. These personas then represent the different needs for which you'll be designing and developing solutions.

Comparing competing solutions by how well they meet needs lets you think outside of the box. Rather than comparing competitors in a defined industry or product category, you can look at any of the means by which people get a job done. Doing so provides important insight into your advantages and disadvantages and what you should focus on.

Finally, define a value proposition based on JTBD. Use the Value Proposition Canvas to align the key jobs that people have with the service you provide or would like to provide.

Regardless of the starting point for innovation in your organization—either with a need or with a technology—you'll need to align around value creation. Then you can involve your whole team in defining what value you'll pursue to get everyone on the same page.

CHAPTER

5

Designing Value

IN THIS CHAPTER, YOU'LL LEARN ABOUT THESE PLAYS:

- How to create a jobs-driven roadmap
- Using job stories to solve specific design problems
- How to architect the structure of a solution
- Testing assumptions directed by JTBD

A software company I once worked for held what were called "hackweeks" once a quarter. This was a time for developers to work on "whatever they wanted," as it was framed. Give engineers time to play around with technology, and they're bound to find the next innovation, or so the theory went.

Hackweek was a big deal for us. Dozens of people organized it, and every developer in the company stopped work to contribute to the effort. It was costly, but we were committed to hackweek. After all, new software offerings come from new development, right?

Here's how it went: small teams formed to cobble together starter projects representing the use of some new technology. At the end of the week, a panel judged the dozens of concepts that emerged, and the winning "solutions" were rewarded.

But in our case, hackweek was like shooting a shotgun in the wrong direction while blindfolded and hoping to hit the target. The result was inevitably a collection of concepts looking for a problem to solve. It was innovation theater at its best.

To be fair, not all hackathons are bad. Some organizations coordinate hackathons with strategic imperatives or with customer needs. And sure, it's also good to flex creative muscles and practice collaboration across teams. But given their cost and imprecision, hackathons are often largely ineffective in producing usable concepts.

The problem is not a lack of ideas—companies are usually swimming in them. Like ours, many organizations have a Darwinistic outlook on innovation: generate more and more ideas, and the best will surely rise to the top. Said another way, when looking for a needle in a haystack, the best approach is rarely to add more hay.

The problem is knowing which ideas to pursue. The goal of innovation activities shouldn't be to collect as many ideas as possible, but instead to get to the right ideas—the ones that matter most to the people you serve.

But more than that, the real challenge is in overcoming the natural forces in organizations that keep good ideas down. Chief among these is uncertainty, a leading deterrent to innovation. New ideas are a gamble for risk-averse managers, even if well-expressed in a high-fidelity prototype.

JTBD provides a way to increase your chances of success by first identifying the right problem to solve. Then JTBD gives you decision-making criteria for moving forward: bet on solutions that address unmet needs to create profitable differentiation.

Focus first on getting the main job done for the individual and fulfilling their needs in relation to the job. From this perspective, hackathons and other idea-generating efforts can be framed by JTBD as both inputs and outputs in terms of how concepts are evaluated.

After understanding the job landscape and defining the value you're going after, you can continue using JTBD thinking to align teams around the design of your solution. Create a roadmap based on your JTBD landscape to set a common direction. Then use job stories to get everyone on the same page and tie local design efforts to the big picture and to architect the solution structure. JTBD can also guide the experiments you conduct to test your team's assumptions.

PLAY ➤ Create a Development Roadmap

At its highest level, a roadmap is a sequence of development events—the relative chronological order in which features and capabilities will be built. Roadmaps serve as a central point of reference for teams to align their efforts. They show the path forward without defining individual tasks.

In the age of Agile and Lean efforts, roadmaps have gotten a bad reputation. People are quick to point out—and rightfully so—that long-term plans inevitably fail: priorities change, unforeseen challenges arise, and timelines slip. The solution, they might argue, is to have no long-term plans and to work on short initiatives with the flexibility to change as needed.

But while providing decision-making power to local development teams makes sense, overall alignment is still needed. An alternative way of viewing roadmaps is to see them not as a definitive project plan, but as a vision of how you'll create an offering that customers will value. Roadmaps are not unchanging predictions of future activity, but a way to provide transparency for the sequence of steps your team will take to design solutions.

The information in a roadmap helps the entire organization get aligned, not just developers. It's a strategic communication tool reflecting intention and direction. More importantly, road mapping isn't just about the artifact: it's about getting a common understanding of where you're headed. In this sense, the roadmap occupies the space between the vision and detailed project planning.

JTBD can help create roadmaps that focus on the value that the organization intends to create and deliver for customers. The trick is to get the right problem to solve. Use the insights from your JTBD investigation to formulate roadmaps that are grounded in real customer need.

Mapping the Road Ahead

For a concrete approach to road mapping, I recommend the book *Product Roadmaps Relaunched* by C. Todd Lombardo, Bruce McCarthy, Evan Ryan, and Michael Connors.[1] In it, the authors clearly articulate the steps to creating meaningful product roadmaps.

JTBD plays a key role in aligning to customer needs, as the authors write: "We recommend starting with the chunks of value you intend to deliver that will build up over time to accomplish your visions. Often this is a set of high-level customer needs, problems, or jobs to be done."

1. C. Todd Lombardo, Bruce McCarthy, Evan Ryan, and Michael Connors, *Product Roadmaps Relaunched* (Sebastopol, CA: O'Reilly, 2017), p. 13.

Their approach breaks down the four key elements of a good product roadmap:

- **Product vision:** The vision outlines how your customers will benefit from your offering. How will the job performer benefit from the solution? What will getting the job done look like after the solution is in place?
- **Business objectives:** A roadmap must be aligned with the organization's strategy and objectives. The goals of the business are important for measuring progress.
- **Timefames:** Rather than committing to specific dates, good roadmaps sequence work and set broad timelines for completion.
- **Themes:** These are the key problems that customers face when completing a job, or clusters of needs that align to the overall solution to be created. JTBD helps frame the themes of your roadmap in particular.

Figure 5.1 shows an example from their book of a basic roadmap overview for a fictional company, The Wombatter Hose, illustrating these main components. Note the disclaimer, as well, indicating that the roadmap is subject to change.

Putting it all together, the process for creating a JTBD-driven roadmap can be broken down into four phases.

THE WOMBATTER Hose

1 PRODUCT VISION
Perfecting American lawns and landscapes by perfecting water delivery

2	H1'17	H2'17	2018	Future
3	**Indestructible Hose** Objectives: • Increase unit sales • Decrease number of returns • Decrease overall defects 4	**Delicate Flower Management** Objective: • Double ASP	**Putting Green Evenness for Lawns**	**Infinite Extensibility**
		Severe Weather Handling Objective: • NE Expansion	**Extended Reach**	**Fertilizer Delivery**

5 **Updated 3/30/17, subject to change without notice.**

1 **Product Vision**
Our garden hose exists to help American consumers pursue the perfect landscapes they so seem to crave. The product vision directly reflects this in compact form, providing an effective framing for everything that comes after.

2 **Timeframes**
Our garden hose roadmap provides for wide half- and whole-year timeframes to ensure the team has the latitude to explore the best ways to solve customer problems.

3 **Themes**
The key problems customers face when watering their landscapes form the themes in the timetable at the heart of the roadmap.

4 **Business Objectives**
Each garden hose theme has one or a few objectives, each of them measuring the business improvement hoped for from solving the customer problems expressed in the theme.

5 **Disclaimer**
A simple date and "subject to change" notification at the bottom of the timetable is sufficient for the limited audience of this roadmap.

FIGURE 5.1 An example of the main components of a roadmap from the book *Product Roadmaps Relaunched.*[2]

2. Image used with permission from the authors.

STEP 1 ➤ Define the solution direction.

Define the various elements of your overall product strategy to get agreement on how you'll be using them. In addition to your solution vision, also define the following together with the team:

- **Mission:** What are your business intentions? The mission is about what your organization wants to ultimately achieve.
- **Values:** What are your beliefs and ideals? What is the philosophy of your organization and solution? Values define the philosophy of the team and what it believes.
- **Business objectives:** What are the specific goals your offerings will accomplish for the organization? Frame these in terms of outcomes, not outputs.

STEP 2 ➤ Determine the customer needs to pursue.

Next, decide on the customer needs to pursue. Here, the authors of *Product Roadmaps Relaunched* stress the importance of grounding the roadmap in actual customer need. JTBD is central to this step. They write:

> Identifying customer needs is the most important aspect of your roadmapping process. Roadmaps should be about expressing those customer needs. Therefore, most items on your roadmap will derive from a job the customer needs to accomplish or a problem the customer must solve.

As outlined in Chapter 2, "Core Concepts of JTBD," needs are hierarchical—from high-level aspirations to main jobs and sub-jobs to micro-jobs. Figure out the top-level jobs to explore and then drill down into the specific themes to target.

The "value themes," as they are called, might come right from the job map. Locate the areas of highest underserved needs and use those stages as the categories of your roadmap themes. Or you can cluster needs to form themes that don't necessarily follow the chronology of the job map. The important point is to ground the division of the roadmap in real-world observations of the customer's job to be done and align the timeline to it.

STEP 3 ➤ Set a timeline.

Next, create a sequence of value themes that your team will work toward. Timelines can be absolute, relative, or a mix of both. Absolute timelines with specific dates carry the risk of changing, which, in turn, can cause confusion or missed expectations.

Relative timelines give more flexibility but still provide insight into what's coming and why. There are various terms to use, but the timeline is often broken into three phases for near-term, mid-term, and long-term. Examples include "now, later, future" or "going, next, later" or something similar. Find what works best for you.

STEP 4 ➤ Align the development effort to the roadmap.

Finally, conceptualize specific solutions to design and create. Use job stories to tie the overall project intent to customer needs, outlined in the next section. Then conceptualize solutions around getting the entire job done or the parts of it determined to be most strategically relevant to your business.

After a roadmap is created, you may then need detailed project plans to track progress. A simple Kanban board can serve that purpose in many cases. Or, for more complex software development efforts, tracking software may be needed. In Agile efforts, epic planning and then sprint planning come after you have an overall roadmap.

Tying the overall plan to customer needs gives the design and development teams the feeling that they are building something that matters to customers. Staying focused on customer needs helps avoid building things your customers don't want. The nature of a job stays the same, even as features may shift. Grounding the roadmap in JTBD ensures that both its longevity and ability to absorb will change.

LEARN MORE ABOUT THIS PLAY:
CREATE A DEVELOPMENT ROADMAP

C. Todd Lombardo, Bruce McCarthy, Evan Ryan, and Michael Connors. *Product Roadmaps Relaunched* (Sebastopol, CA: O'Reilly, 2017).

> This book distills a wealth of practical information into a compact guide on roadmapping. The authors go to great lengths to provide numerous examples and stories from real-world cases. They use a realistic, modern approach for creating a roadmap that is driven, in part, by JTBD. A special thanks goes out to the authors for granting permission to use an image from their book.

PLAY ➤ Align Teams to Job Stories

Agile development enables teams and organizations to work in a flexible way. The approach started in software development, but has spread to other domains, including government and the military. The principles of Agile development can apply to just about any field.

A key part of Agile is to break down efforts into individual units of work. *User stories* are short descriptions of features and functionality written from the perspective of the end user. Teams can focus on only a small part of the whole and make progress in a controlled way.

User stories are commonly written in a three-part format. The first element indicates a user's role in the system. The second points to a capability that enables the person to get a task done. The last part often describes a benefit or reason for using the capability.

Although specific styles can vary, a typical user story resembles something like the following:

> As a <role> I can <capability>, so that <benefit>

Examples of user stories in this format include:

- As a system admin, I can specify files or folders to back up based on file size, date created, and date modified.
- As a user, I can indicate folders not to back up so that my drive isn't filled up with things I don't need to be saved.
- As a user, I want to update the name of a document so that I can categorize it.

For any given system, there may be hundreds of user stories. Some can be quite granular, such as describing a single button and why a user would click it. Stories are then organized into a backlog or repository of functionality to be built. Teams break off logical groups of user stories in sprints or two- to four-week cycles of work.

Job Stories

Although user stories are good for breaking down work, they typically fail to connect the solution being built with user needs. They lack an indication of *why* someone would behave in a certain way and what they need to get a job done. In fact, often user stories are derived from the capability being built, not from observing actual behavior.

Job stories are an alternative to user stories. They follow the tradition of breaking down efforts into smaller pieces, but through the JTBD lens. The technique was first pioneered by the product development team at Intercom, a leading marketing communications solution. They wanted to avoid leading designers with a preconceived solution, as well as tying development to the company vision and strategy.

Paul Adams, an Intercom product manager, wrote about job stories for the first time, saying: "We frame every design problem in a Job, focusing on the triggering event or situation, the motivation and goal, and the intended outcome."[3]

3. Paul Adams, "The Dribbblisation of Design," *Inside Intercom* (blog), September 2013.

As a result, their job story format also has three parts. But instead of focusing on a generic role, like a "user" or an "admin," job stories begin with a highlight on the situation and context, not the individual:

> *When [situation], I want to [motivation], so I can [expected outcome].*

Examples of job stories include:

- When an important new customer signs up, I want to be notified so that I can start a conversation with that person.
- When I visit someone's profile page, I want to see how many posts they have in each topic so that I have an understanding of where they have the most knowledge.
- When I have used the application multiple times, I get nudged to contribute so that I am encouraged to participate.

JTBD author and leader Alan Klement has done the most work refining the job story format.[4] He believes that adding more information about the circumstances shows causality better. Focusing on the context shifts attention from a persona to the situation. Klement advises that you avoid writing vague situations, but instead be as specific as possible.

For instance, consider these three possible situations for the first element of job stories:

- When I'm hungry…
- When I'm lost…
- When I want to check my email…

4. See for example: Alan Klement, "Replacing the User Story with the Job Story," *JTBD.info* (blog), November 12, 2013.

Instead, Klement recommends describing the circumstances in rich detail:

- When I'm hungry, running late to get somewhere, not sure when I'm going to eat again, and worried that I'll soon be tired and irritable from hunger…
- When I'm lost in a city that I've never been to, don't know the local language, and am worried that I'll be wasting my time in places I don't want to be in…
- When I want to check my email, but don't want anyone around me to know I'm checking my email because they'll think I'm being rude…

Each of these example situations provides more context for designing an appropriate solution.

Working with Job Stories

Job stories are modular, giving designers and developers the flexibility to solve problems in alternative ways. Job stories are grounded in real-world insight, and they are more powerful than user stories in guiding solutions. But creating job stories is more free-form than other JTBD techniques. Still, there are patterns that you can follow. Using the elements from Chapter 2, I suggest the following structure for job stories:

> *When I [circumstance + job stage/step], I want to [micro-job], so I can [need].*

Examples:

- When I am one of the top posters while updating my social media feeds daily, I want it to show on my profile so that I can increase recognition as an expert on the subject.
- When I run out of materials needed while completing an art project, I want to find alternative materials so that I can maximize the number of uses of my current supplies.

- When preparing for my commute and running late, I want to know the current weather along my journey so that I can minimize the chance of arriving wet.

Consider the last example. The first element combines information about the circumstances (*running late*) of getting the main job done (*commute to work*) within a stage of the process (*prepare for commute*).

The second element points to an even smaller step or micro-job (*check forecast*). It should be formulated without reference to specific technology, but should be specific enough for designers and developers to create a specific capability.

Finally, the last element can be taken right from your list of needs. In this case, the job performer (*commuter*) wants to avoid showing up to the office wet (*minimize the chance of arriving at work wet*). You can leverage the elements your JTBD landscape already uncovered in research directly in the formulation of the job story statements.

In researching this book, I've come across various alternative approaches to formulating job stories. Andrea Hill, a prominent advocate of JTBD on social media, suggests a slightly different approach. She sees the middle element pointing directly to a feature or solution of some kind, thus explicitly crossing from the problem space into the solution space. Her basic format is as follows:

> *When I [circumstance], I want to [solution capability], so I can [need].*

A job story for the previous example of commuting to work might then look like this:

> When I'm preparing to commute to work, I want to have weather forecast notifications pushed to my phone, so I can minimize the chance of arriving wet.

Steph Troeph, research and JTBD instructor in the UK, approaches job stories in yet another way. She thinks of them with this formula:

> *When I [circumstance], I want to [job], so that [benefit a solution offers].*

Regardless of your interpretation, the key is to find a consistent structure and stick with it. The form you end up with needs to be appropriate to your team and your situation.

Job Stories in Action

Ultimately, job stories tie a local design and development effort to a broader JTBD framework. Because the format of job stories includes contextual details, they are portable. In other words, a job story should make sense without having to know the larger JTBD landscape or job map. As a result, job stories have a more "plug-and-play" versatility that is often required for Agile designs and development teams.

For instance, Agile planners can manage a backlog of job stories much in the same way that they would manage user stories. If a given sprint gets slowed down or changes direction, stories not addressed can be carried over to the next sprint. Having a smaller, self-contained description of the smaller job to be done has advantages during the design and development phases.

But to be clear: I have found that job stories typically *do not* replace user stories for development completely. Instead, job stories guide and frame the conceptualization of a solution rather than track implementation. They serve best as a design tool to create or determine concept direction and design. Developers and engineers will likely still need user stories to measure the burndown rate and overall progress.

Your job map provides an overall orientation to your JTBD landscape and allows you to zero in on a specific area for design and development. A roadmap gives you a high-level sequence of development with the rationale for planning activities. Job stories are more specific and guide the local design and development of features and capabilities.

Follow these steps to create job stories based on your JTBD research:

STEP 1 ➤ Understand job stages and circumstances.
Base the relevant jobs and circumstances on previous interviews and observations. For each area of development in your solution, consider the steps in the main job. Then drill down and list the smaller and smaller steps as micro-jobs, using the rules of formulating JTBD. Also identify the circumstances that apply to that part of the main job in particular.

Depending on the depth of your prior research and how well you and your team understand the job, you may not need to do more research to create and validate job stories. It's never a bad idea to speak with people again and drill down on specific problems and objectives they have. During additional interviews, ask "how?" until you get more granular in understanding of subgoals and objectives.

STEP 2 ➤ Formulate job stories.
As a team, write job stories that are specific to your design and development effort. Decide on a consistent format for the job stories and stick to it.

Strive to come up with unique, mutually exclusive stories that target specific jobs and circumstances. Avoid redundancy. For instance, in the previous example, you probably don't need separate stories for commuting by train versus commuting by car. Develop the job stories that matter the most and focus on a limited set. You may end up with anywhere from three to eight job stories per project or sprint.

STEP 3 ➤ Solve for the job stories.
Make the job stories visible to the entire team to solve for the underlying need. For instance, post a relevant list of job stories in a brainstorming session for everyone to see. Or list job stories at the

beginning of a design critique so that the team has context for making comments. Use JTBD to guide design and development decisions.

It's also possible to then use the job stories to review the appropriateness of your solutions. First, the design team can use the job stories relevant to a project as heuristics. They should constantly ask if their designs are meeting the user's goals set out in the job stories.

Then you can test solutions with users against the job stories. Show users your solutions (e.g., as a mock-up or prototype) and ask them how well each addresses the job stories. This can be done in an interview-style fashion or with a survey. The job stories ultimately become a measure for success of the designs before anything is built.

Job stories let you take a step back and look at the context of the job while designing a product or service. In this respect, job stories fill an important gap between the observations of customers and solution development, connecting insights into customer needs to individual features and development efforts.

Related Approaches: Need Statements

Design thinking is a broad framework for creative problem solving. It is rooted in human-centered methods that seek to develop deep empathy for people and then to devise solutions that meet their needs. In design thinking, it is important to define the problem to solve before generating options for solutions.

One technique to encapsulate insights from research is to generate *need statements*, greatly resembling job stories in form. But these statements differ from "needs," as defined in Chapter 2, in that need statements in design thinking are not specifically limited to the outcomes of getting a main job done, and they can be aspirational in nature.

Need statements in design thinking also tend to be much more focused on a persona or an individual rather than the circumstances. For instance, writing for the Nielsen Norman Group, Sarah Gibbons refers to need statements representing a point-of-view for the user of a system:[5] "A user need statement is an actionable problem statement used to summarize who a particular user is, the user's need, and why the need is important to that user."

Like job stories, need statements have three components: a user, a need, and a goal. The *user* corresponds to a goal-based persona based on research (as outlined in Chapter 4, "Defining Value"). A *need* is expressed independent of a feature or technology. The *goal* is the result of meeting the need. Gibbons provides an example:

> Alieda, a multitasking, tech-savvy mother of two, needs to quickly and confidently compare options without leaving her comfort zone in order to spend more time doing the things that really matter.

Note that the insight at the end of this statement, "doing the things that really matter," is very broad and hard to measure. Job stories, on the other hand, favor a more specific context and outcome. For instance, rewriting the above example through the lens of job stories might yield something like the following:

> When I'm multitasking and in a rush, I need a familiar way to quickly and confidently compare options so that I can minimize the time spent on finding a solution.

Like need statements in design thinking, job stories also avoid the mention of features or technology. Yet, they are much more specific to a given job and its context. While both a need statement from design thinking and a job story can feed into the creative generation of

5. Sarah Gibbons, "User Need Statements: The 'Define' Stage in Design Thinking," (*article, Nielsen Norman Group, March 24, 2019*).

solutions, job stories will provide more direct guidance without prescribing a solution.

But the definition of a *need* in design thinking can vary greatly. For instance, IBM's Enterprise Design Thinking approach also includes guidelines for generating statements.[6] Not surprisingly, there are three parts: a user, a need, and a benefit. Here's an example from the IBM site:

> A developer needs a way to make sense of minimal design so that they can prototype faster.

This example is much more specific than Gibbons's approach, yet still avoids mentioning a specific solution. There are no aspirational elements, such as "pursuing lifelong dreams," sometimes found elsewhere in design thinking. IBM's approach to need statements is closer to the job story approach, but is also light on describing the circumstances of use.

In some sense, the differences between job stories—even with the variations in format—and need statements points to a key distinction between JTBD and design thinking. The former focuses much more on the circumstances than the person's state of mind or psychology. Where design thinking seeks to gain empathy for the individual as a starting point, JTBD seeks to understand the circumstances of accomplishing an objective before factoring in emotional and personal aspects.

6. See : "Needs Statements" in IBM's *Enterprise Design Thinking Toolkit* (August 2018), available online.

LEARN MORE ABOUT THIS PLAY:

ALIGN TEAMS TO JOB STORIES

Alan Klement, "Replacing the User Story with the Job Story," *JTBD.info* (blog), November 12, 2013; "5 Tips for Writing a Job Story," *JTBD.info* (blog), November 12, 2013; "Designing Features Using Job Stories," *Inside Intercom* (blog), 2013.

> Klement has done the most extensive work to develop the job story technique. These three articles outline the basis for creating them. The technique has evolved slightly, but Klement points clearly to how he's updated his approach. Klement and others have posted widely about their use for development efforts, but start with these resources.

Maxim van de Keuken, "Using Job Stories and Jobs-to-be-Done in Software Requirements Engineering," (thesis, Utrecht University, 2017).

> This thesis project offers a detailed investigation of how job stories are applied to date. After illustrating the history of job stories, Van de Keuken presents the results of his original research variations in application of job stories as seen in practice. This work contributes greatly to making job stories a more formal part of software requirements engineering.

PLAY ➤ Architect the Solution

JTBD not only informs upfront strategic decisions, but it also guides the design of a given solution. In particular, product architecture can be derived from JTBD research.

"Architecture," in this sense, refers to the foundational structure of how various components of a solution come together conceptually. This is not the technical architecture, nor the organization of interface components. Rather, it's about the underlying composition of a solution—how the parts are organized. Ideally, this organization comes from patterns identified in the job to be done. Matching the model of the system to the model of the job ensures better comprehension, better usability, and ultimately better product-market fit.

Architecting a solution recalls Jesse James Garret's infamous model for user experience design, which has five layers (see Figure 5.2).[7] In the middle is "Structure," or how the pieces of a product or service fit together conceptually. Prior to that, however, the product provider needs to determine the strategy and scope. After the structure is determined, the skeleton and surface of the interface are created. Overall, design moves from the abstract to the concrete with these layers.

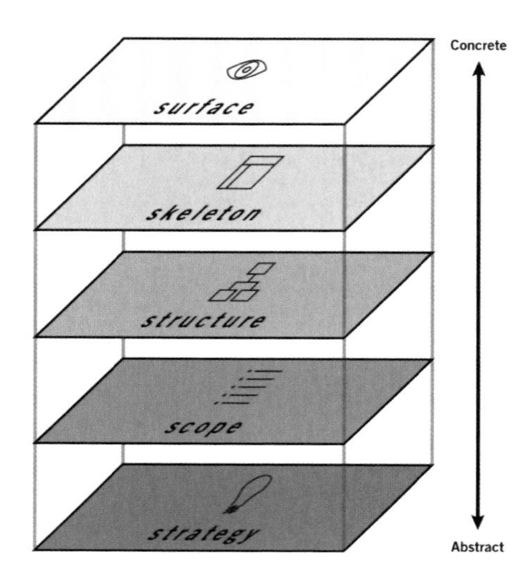

FIGURE 5.2 Garret's model of experience design shows conceptual layers below the surface.

Paul Adams, head of product at Intercom, discusses the importance of getting the right structure in his article "The Dribbblisation of Design," the same article that introduced the concept of job stories.[8] The archi-

7. For a detailed description of the diagram, see: Jesse James Garret, *Elements of User Experience* (San Francisco: New Riders, 2002).
8. Paul Adams, "The Dribbblisation of Design," *Inside Intercom* (blog), September 2013.

tecture of a solution should not be based on technology, but on jobs to be done. He writes:

> After the mission and vision is the product architecture. Not the technical architecture, rather the components of your product and how they relate to one another. The system.
>
> This gives us clarity. We can map this Job to the mission and prioritize it appropriately. It ensures that we are constantly thinking about all layers of design. We can see what components in our system are part of this Job and the necessary relationships and interactions required to facilitate it.

As an example, Adams points to the architecture of Facebook, shown in Figure 5.3. For each component of the system, there is a corresponding job step such as *compose a message*, *send a message*, and *create a profile*. These become the basis of the solution organization independent of technical or interface concerns.

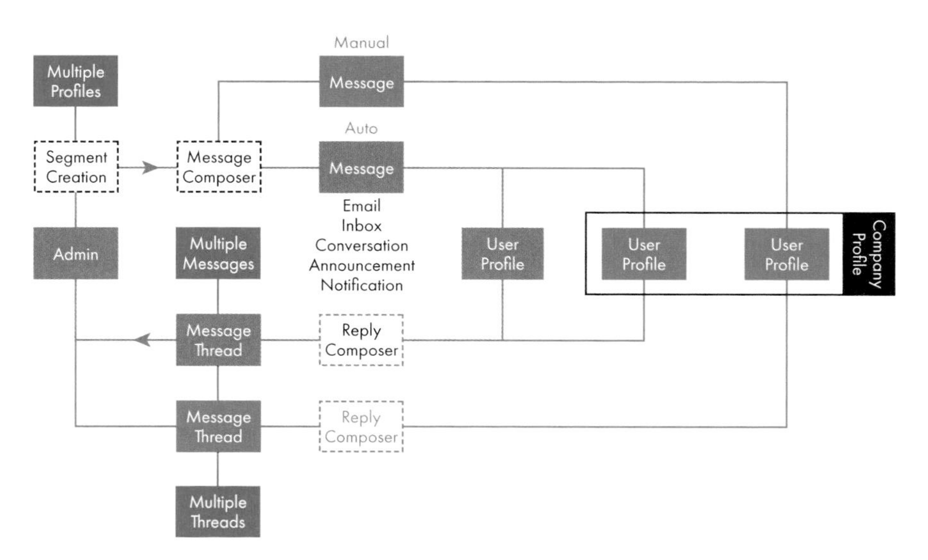

FIGURE 5.3 An example of a product architecture.

Modeling Systems

Solution architecture is abstract. Models make it tangible. Creating a model of the solution structure ties your understanding of customers to the overall conceptual design. Even a simple diagram, such as the one shown previously in Figure 5.3, can help get a team on the same page.

Think about the system model like the floor plan of a building. You don't see it when entering a building or a room, but it's there, framing your experience as you move around inside. Instead of walls and floors dividing the physical space, in solution architecture, categories create the divisions of the model. These categories ideally come from jobs to be done. Aligning the basis of your model with jobs helps ensure that the solution will match the mental model of the eventual user.

Leverage JTBD at the beginning of the design to inform your solution architecture. The key is to ground design in real-world observations. Good solutions have an inherent logic to their structure that mirrors the user's mental model of their job to be done (see Figure 5.4).

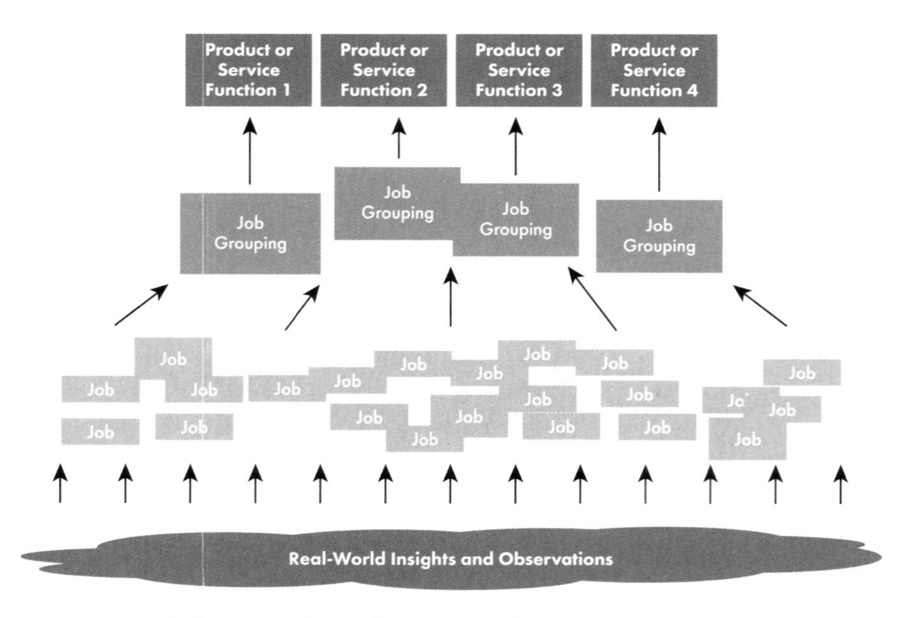

FIGURE 5.4 Architecting solutions from JTBD is a bottom-up process.

USER ENVIRONMENT DESIGN (UED)

In their landmark book *Contextual Design,* Beyer and Holtzblatt propose a specific technique for structuring products called *User Environment Design* (UED).[9] Although specifically framed for software design, UED can be applied to structuring any solution, including service design and even industrial design efforts.

UED is based on an understanding of what the authors describe as the "work" that users are trying to get done. Although they don't use the phrase directly, this notion overlaps with jobs to be done. Beyer and Holtzblatt show that the structure of a product or service should mirror the user's work, rather than technology. If you support the user's work as best as possible, then you'll have the best chance of adoption.

As an example of UED, Beyer and Holtzblatt point to an application that manages email, as seen in Figure 5.5. The core messaging system is represented in the green boxes in the middle. Above that is a distinct area of the product for managing setup and preferences. Below the core is a set of functions that support the work of creating and sending messages.

These divisions are structural issues, not interface issues. In later design phases, the UI may reflect some of these categories and labels, but the surface can have more detail. So the system structure design at this point is abstract and not about concrete design choices on the surface. The point is that mirroring the architecture to jobs to be done helps ensure that your overall solution will be more useful and usable.

After you observe users in the context of getting a job done, identify key patterns based on needs. Then arrange these patterns into a logical model within the scope of your solution. After that, your team can design both the technical architecture and the user interface.

9. Hugh Beyer and Karen Holtzblatt, *Contextual Design* (New York: Morgan Kaufmann, 1998).

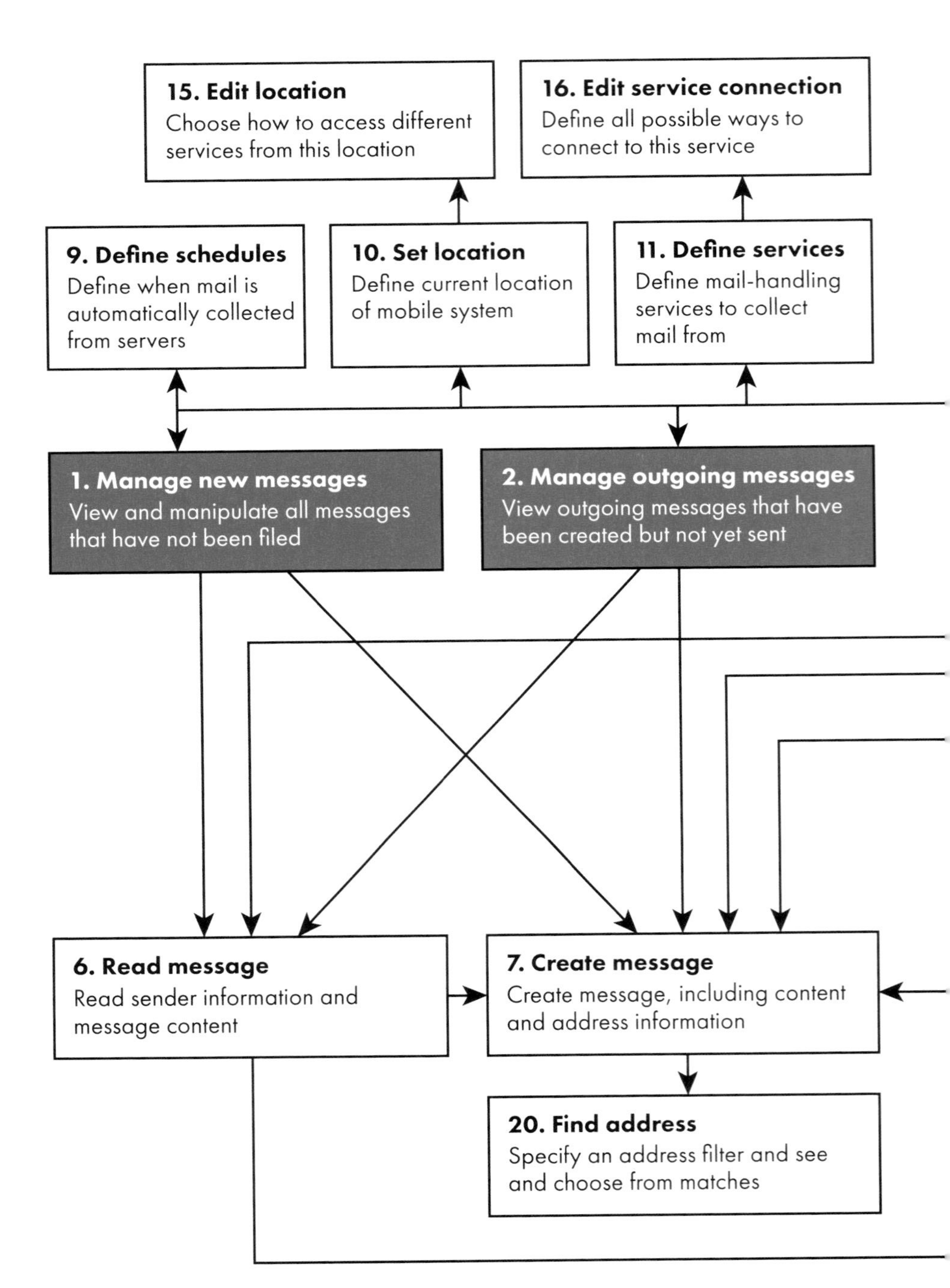

FIGURE 5.5 An example of a solution architecture for an email system.

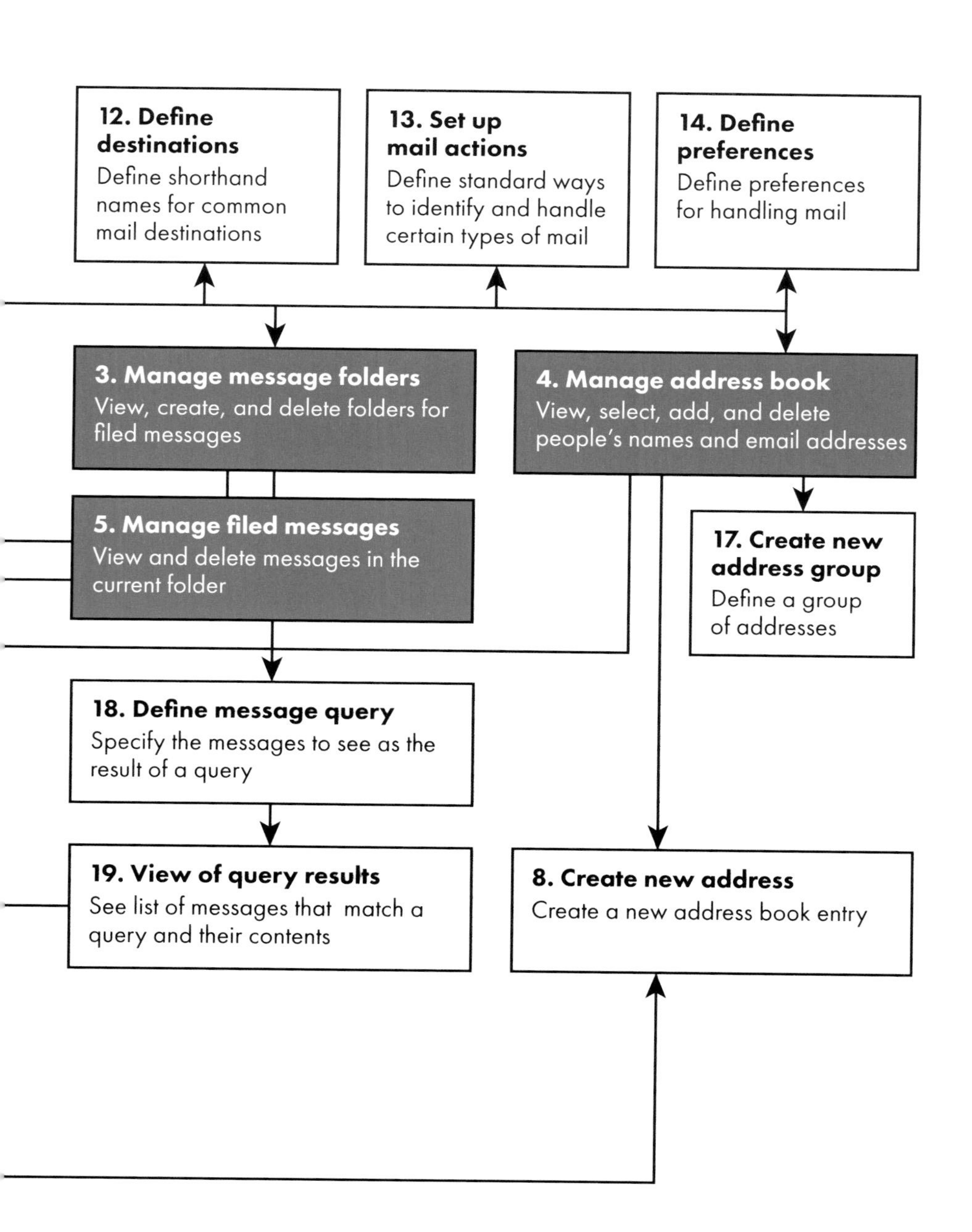

12. Define destinations
Define shorthand names for common mail destinations
13. Set up mail actions
Define standard ways to identify and handle certain types of mail
14. Define preferences
Define preferences for handling mail
3. Manage message folders
View, create, and delete folders for filed messages
4. Manage address book
View, select, add, and delete people's names and email addresses
5. Manage filed messages
View and delete messages in the current folder
17. Create new address group
Define a group of addresses
18. Define message query
Specify the messages to see as the result of a query
19. View of query results
See list of messages that match a query and their contents
8. Create new address
Create a new address book entry

STEP 1 ➤ Understand users and their work.

As usual, begin with a deep understanding of individuals and their jobs to be done. Job interviews provide the foundation for modeling the user environment. If you've already conducted research, then extract as many micro-jobs as you can. Write them on individual cards or individual boxes.

STEP 2 ➤ Identify focus areas.

Unlike a job map, which is chronological, a model of the solution will typically have no temporal component. Start by clustering micro-jobs into logical groups, called *focus areas*. Since you are targeting a specific solution at this point, consider the end user's mental model in getting work done. What categories are relevant in getting the main job done? How might an end user conceptualize their work?

Label each cluster with a simple title that follows the rules for formulating JTBD to the degree possible. Your solution also may require elements in your model that go beyond pure JTBD research. For instance, the solution might require an administration area, which represents a consumption job. The aim is to organize the components of the system around user objectives.

STEP 3 ➤ Model the solution structure.

Finally, arrange the focus areas into a network diagram. Show relationships between them with arrows and lines. The diagram is like a floor plan of a house, reflecting the major rooms and the key pathways between them. Include a list of specific functions within each focus area.

Focusing on users' "work" in modeling the user environment, as outlined in *Contextual Design*, closely resembles JTBD thinking. Other approaches to the type of modeling previously discussed also have overlap with JTBD. Regardless of the particular approach you choose to follow, modeling product and service solution architectures is a bottom-up approach. First, observe the individual's job to be done and understand that.

Related Approach: Web Navigation Design

The term *mental model* refers to someone's thought process about how the world works—their frame of reality. Mental models allow you to predict how things work. They are cognitive constructs built on beliefs, assumptions, and past experiences.

The mental model the person has of the system is framed by that system. If you explore their mental model independent of a solution, then you can break out of the system frame. You can discover aspects of how a person thinks that have nothing to do with the system, but everything to do with how that person accomplishes their intent.

Author and UX researcher Indi Young has developed a specific approach for mapping mental models.[10] Her diagramming technique seeks to understand and visualize people's intent and purpose in a given domain. These models can be directly used to derive a website navigation that best matches the understanding of the website a user might have.

The hierarchical nature of mental model diagrams makes them particularly relevant for the practice of information architecture. The process can be described as grounded: a bottom-up approach starting with summaries of how people describe their reasoning, reactions, and guiding principles. Then it's a matter of successively grouping information into higher level categories.

Figure 5.6 shows just one section of a mental model diagram—this shows the job of seeing a movie at the cinema. It reflects a hierarchy of goals and intent.

10. See her landmark book: Indi Young, *Mental Models* (New York: Rosenfeld Media, 2008).

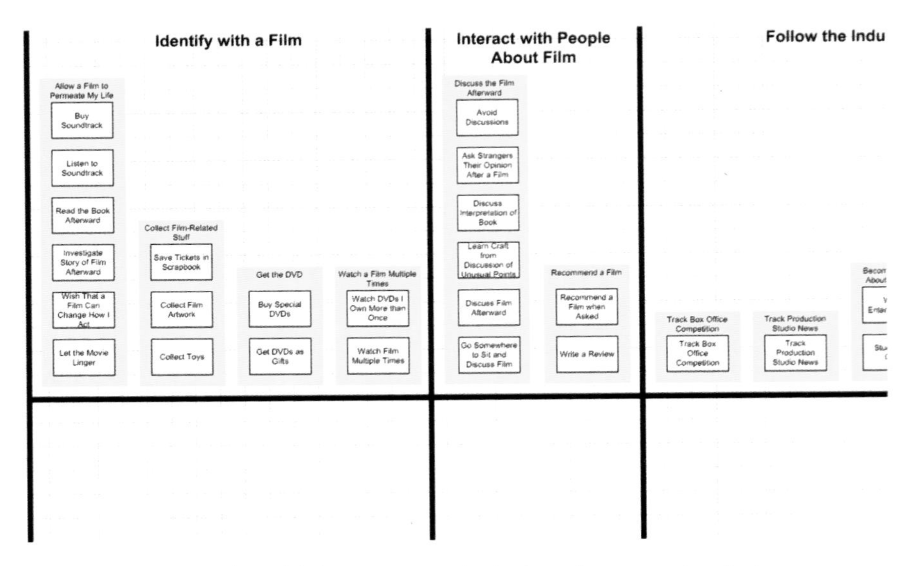

FIGURE 5.6 Mental model diagrams provide the basis for navigation design.

The result is a categorization that matches the actual mental model of the people you serve and reflects the vocabulary that people have used in interviews. Web designers, for example, can then use this scheme as the basis for navigation. This greatly improves usability of the navigation and ensures its longevity as well.

Young describes the process of deriving structure and mapping it to navigation in detail, resembling the steps in UED, outlined previously. Figure 5.7 shows how mental spaces can be grouped into categories that then serve as the main navigation for a website. You should refine categories and labels. Then start to represent the navigation in wireframes to indicate their position and interaction.

From a JTBD perspective, there are two approaches to categorization of a domain that are related. First, you could look at the individual steps of a job map to find logical groupings. However, since a job map is by nature chronological, you'll likely end up with categories that reflect the phases of a process.

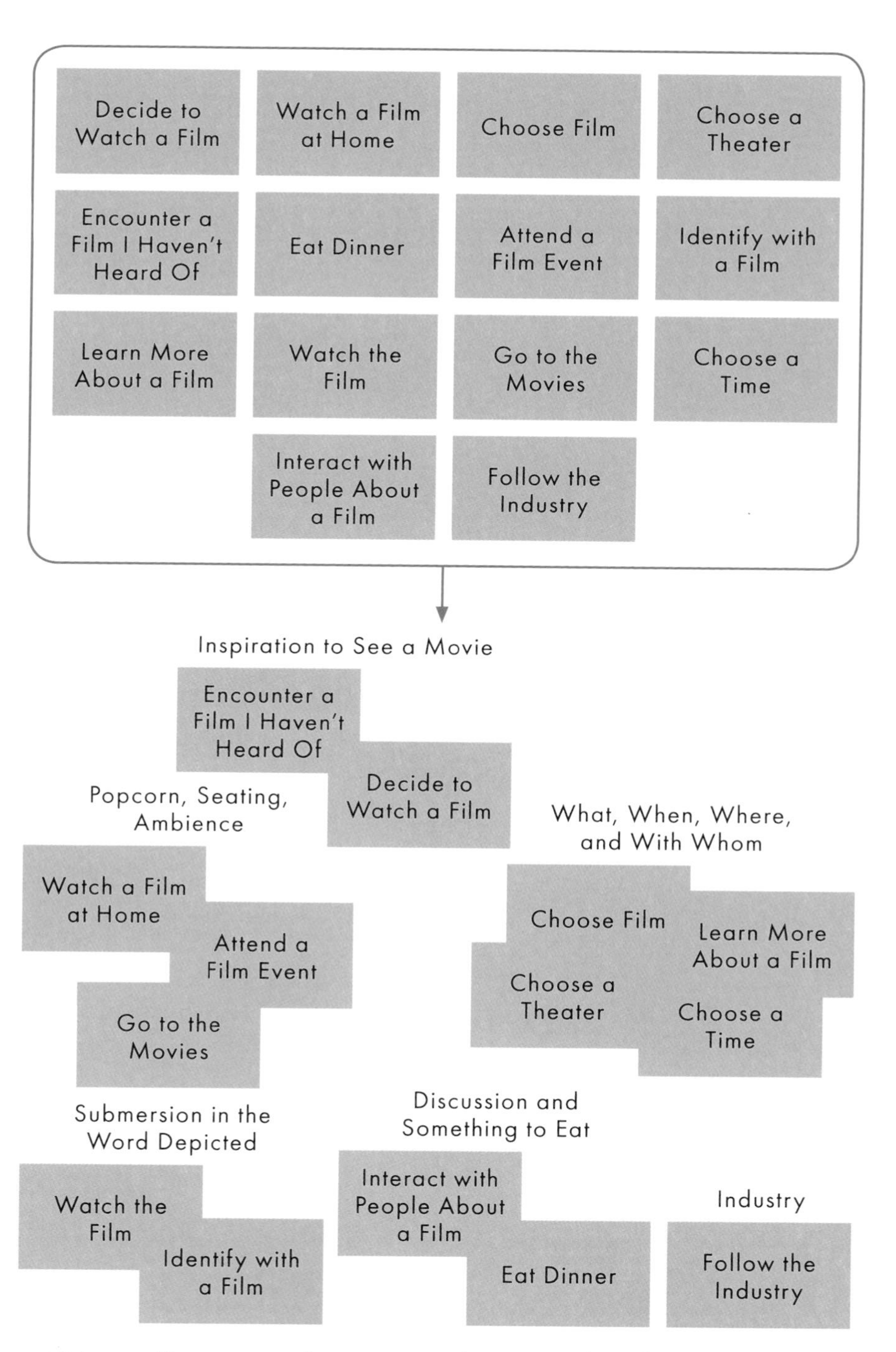

FIGURE 5.7 Cluster user goals to come up with categories for website navigation.

For instance, if you cluster the stages in a job map for growing a retirement portfolio, you may end up with categories such as "get started," "create a plan," "monitor growth," and "make adjustments."

In terms of website navigation, these could be viable navigation options. But that system of labels might not cover all of the content topics available on your site. There could be sections on how to include family members in your decision-making process or about related financial topics, such as buying a home or financing college.

Another option is to categorize need statements for the main job. The process is similar to the technique shown in Figure 5.7. From the bottom up, create logical categories of need statements and label those categories. If needed, cluster the categories into larger buckets and use those as inspiration for a main navigation. You'll likely have to modify the labels and options, but at a minimum, it's possible to base a website navigation on JTBD with this approach.

LEARN MORE ABOUT THIS PLAY:
ARCHITECT THE SOLUTION

Hugh Beyer and Karen Holtzblatt, *Contextual Design*. (San Francisco: Morgan Kaufmann, 1998).

> *Contextual Design* presents a complete approach for going from real-world observations to designing software interfaces. Their notion of focusing on user "work" is extremely close to the concepts of JTBD. Although their approach is specifically geared toward software design, the approach can inform the creation of just about any solution type.

Indi Young, "Structure Derivation," Chap. 13 in *Mental Models* (New York: Rosenfeld Media, 2008).

> This chapter offers one of the best explanations of tying research insights to product architecture. It's a bottom-up process: you can cluster observations multiple times to derive high-level categories. Young also discussed how to label parts of the resulting architectural model.

PLAY ➤ Test Hypotheses with JTBD

Innovation brings uncertainty. No inventor can predict market success. Introducing new offerings to the market is risky because the consumers ultimately decide to adopt an innovation or not, regardless of the inherent benefits of the solution.

Just consider the launch of the Segway, the famous self-propelled one-person scooter. The invention itself worked great and was very compelling to investors. But ultimately, the market rejected it as a viable means of daily transportation. Getting the functional job done alone is not enough. Emotional and social factors, as well as the circumstances, come into play when devising a solution that people really want.

However, had the manufacturers done some simple experiments in advance, they might have seen this reaction and adjusted before it was too late. Waiting until a product or service is built to test it takes too long and is wasteful. Instead, it's possible to conduct experiments before anything is created. Your aim is to learn about product-market fit as soon as possible and adjust accordingly during the design and development phases.

Leveraging JTBD in Experiments

In their book, *The Customer-Driven Playbook*, authors Lowdermilk and Rich show how JTBD can be used to formulate testable hypotheses throughout the product design and development lifecycle.[11] JTBD ties the understanding of the underlying problem directly to the solution by providing a common thread and a common language.

The authors' approach has four stages called the *Hypothesis Progression Framework* (HPF). The first two stages—understanding the *customer* and the *problem*—fall under "customer development. These roughly

11. Travis Lowdermilk and Jessica Rich, *The Customer-Driven Playbook* (Sebastopol, CA: O'Reilly, 2017).

correspond to problem space understanding. The "product development" stages include the *concept* and then feature *development*. Overall, the divisions in their framework resemble the four Ds (discover, define, design, deliver) used to organize this book, with some differences.

The HPF allows you to test your assumptions at any stage of development. At each stage, the authors recommend that you should develop and test your hypothesis.

In particular, while creating a solution, the authors recommend a formula for generating hypothesis statements incorporating the JTBD. At the feature level, they propose this format:

> *We believe that [type of customer] will be successful solving [problem] using [feature] while doing [job to be done].*

Create a solution that addresses the unmet needs and test it.

STEP 1 ➤ **Formulate hypotheses.**

Recognize your assumptions at each stage and formulate testable hypotheses. Lowdermilk and Rich have a format for hypotheses for each stage, shown in Table 5.1.

TABLE 5.1 THE HYPOTHESIS PROGRESSION FRAMEWORK (HPF) BY LOWDERMILK AND RICH

CUSTOMER **Who are our customers?**	We believe [type of customers] are motivated to [motivation] when doing [job to be done].
PROBLEM **What problems do they have?**	We believe [type of customers] are frustrated by [job to be done] because of [problem].
CONCEPT **Will this concept solve their problem?**	We believe that [concept] will solve [problem] and be valuable to [customers] while doing [job to be done]. We will know this to be true when we see [criteria].
FEATURE **Can they use this feature?**	We believe that [type of customers] will be successful solving [problem] using [feature] while doing [job to be done]. We will know they were successful when we see [criteria].

As the name suggests, your aim isn't just to experiment at one stage, but throughout the product and service development. Keeping JTBD consistent for each of the four different hypothesis types helps you ensure that the team will ground their learnings in relevant findings. This method ensures consistent, continual learning. Your aim is to mitigate risks and increase the chances of adoption.

STEP 2 ➤ **Validate or invalidate hypotheses with experiments.** In the early stages of HPF, interviews and customer visits are good ways to validate your hypotheses. Surveys and analytics can also help prove your beliefs.

Once you have a solution, you can conduct more elaborate experiments of concepts and features. A so-called "minimum viable product" (MVP) can provide a wealth of business insight without having to build or launch anything. Think of a MVP as the shortest path to learning, not as building a product. Eric Ries, author of *The Lean Startup*, which describes a complete approach for business to mimic the experimental behavior of startups, explains in his book:[12]

> The minimum viable product is that version of a new product which allows a team to collect the maximum amount of validated learning about customers with the least effort.... MVP, despite the name, is not about creating minimal products.

Specific approaches to business experimentation within the Lean school of thought include the following:

- **Explanatory video:** Create a video explaining your service and circulate it on the internet. Measure interest via traffic and response rates.

12. Eric Ries, *The Lean Startup* (New York: Crown, 2011).

- **Landing page:** Sometimes called a "fake storefront," you can gauge market interest by measuring traffic and response rates to a simple landing page announcing the future launch of your proposed service.
- **Prototype testing:** Simulate a functioning version of your concept. Test this with potential customers and measure concrete aspects such as task completion and satisfaction.
- **Concierge service:** Start with a manual version of your service. Invite a very limited set of potential customers to sign up and then provide the service manually.
- **Limited product release:** Create a version of your service with only one or two functioning features. Measure the success and appeal of those features.

For instance, Steve Blank, a father of the Lean movement, recounts a previous consulting engagement in a blog post.[13] Instead of building hardware and software products to test an idea, Blank recommended that his team rent the hardware and crunch the numbers by hand. Then they gave the results to potential end consumers to see if they found the service at all useful. No development was needed, and the turn-around time for learning was days, not weeks or months, to build a prototype.

STEP 3 ➤ Make sense of learning and move forward.

After formulating and testing hypotheses at each stage, reflect on what you learned. Determine whether you should hold, change, or kill each component tested.

Collect and store the data from your experiments and extract the parts that serve as evidence to prove or disprove your hypothesis at a given

13. Steve Blank, "An MVP Is Not a Cheaper Product, It's About Smart Learning," (blog), July 22, 2013, https://steveblank.com/2013/07/22/an-mvp-is-not-a-cheaper-product-its-about-smart-learning/

stage. Tag the particularly relevant findings using the parameters from your hypothesis statement, e.g., [problem] or [motivation]. Then create a compelling story of how your investigation impacts the business's course of action.

The goal is to learn, not to develop. If your project team is not open to learning and already believes it knows the right direction, testing hypotheses may not make sense.

JTBD theory helps since it provides a consistent basis for what people are trying to accomplish. The theory itself predicts that people are motivated first and foremost to get a job done. The HPF gives you a clear and structured way to test your assumption at any stage of development.

LEARN MORE ABOUT THIS PLAY: **TEST HYPOTHESES WITH JTBD**

Travis Lowdermilk and Jessica Rich, *The Customer-Driven Playbook* (Sebastopol, CA: O'Reilly, 2017).

> This slim volume is packed with techniques and advice that can be used across product development, from understanding your market to creating solutions that fit. Their approach is skewed toward solution space JTBD, or why people "hire" a solution (a la Clayton Christensen). They provide a rich, complete framework for testing assumptions via hypotheses. JTBD is an element that carries throughout their stages of development.

Ash Maurya, *Running Lean* (Sebastopol, CA: O'Reilly, 2012).

> This book contains a wealth of practical details on how to run Lean experiments. Maurya has laid out a clear path for validating business concepts and product ideas. While he only briefly mentions JTBD in this volume, Maurya uses jobs thinking in his trainings and talks.

For more on business experimentation in general, also see: Eric Ries, *The Lean Startup*, New York: Crown, 2011; Steve Blank, *Four Steps to the Epiphany*, K & S Ranch, 2005; and Michael Schrage, *The Innovator's Hypothesis*, Cambridge: MIT Press, 2014.

Case Study: CarMax

By Jake Mitchell, Principal Product Designer at CarMax

"Could you look at another picture?"

The research participant clicked the "next" arrow and an image of a car for sale appeared on the screen in front of them. The room stayed silent for a moment.

"This is, um, a nice photo," the participant said. "I like that you can see how clean the seats are. That's… nice." The participant clicked the arrow a few more times, pausing every few seconds, trying to come up with something more to say.

It was clear from this research session, and many others like it, that we had hit a wall. I was part of a team at CarMax that was responsible for presenting our used car listings online and ensuring that we were displaying our cars with the best photography possible. The problem was that we had run out of tangible ways to make the photos better.

We were doing all the things we were told to do in today's age of technological disruption: constantly talk to our customers, experiment our way to the best version of the product, and always move fast. But no matter how good we were at embracing these values, none of the methods we were using seemed to be producing results that anyone would consider meaningful. We would bring in people who claimed to be in the market for a used car and try to learn how to improve the photo experience based off the feedback they gave. Then we would implement changes to the experience, only to see little to no noticeable impact to our business metrics.

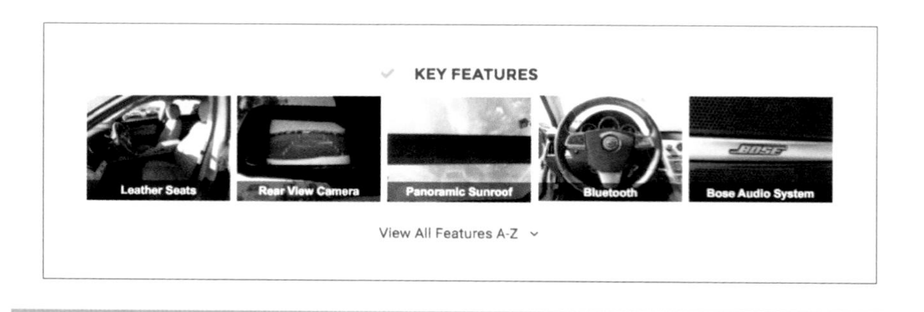

It wasn't until I came across JTBD, or the "Jobs to Be Done" theory that our team began gaining traction. After diving into the material, I realized that we weren't paying enough attention to understanding the underlying needs that would make customers turn to photos. Our focus was purely on how to improve the cosmetic attributes of the photos—i.e., the quality of the photo, the angle, whether it was best to shoot indoor or outdoor, etc. Instead, I realized we should be focusing on how useful the photos were in solving the jobs in people's lives. What we were doing would have been akin to a team of engineers deliberating on what color the fins of an airplane should be instead of figuring out how to make it aerodynamic.

I took this line of thinking back to my team and posed the question to them. "Maybe we're looking at this the wrong way. What are the *jobs* that people are hiring photos for? What are people thinking when they turn to photos?"

Instead of asking people for what they desired, we observed how they used our product and examined the motivations that drove their behaviors. Instead of speaking to anyone who said they had vague plans to purchase a car in the near future, we selectively chose to speak to people who had recently purchased a car. This way, research participants were able to reflect on the choices they had made during the shopping process and show us their thinking behind it, instead of performing hypothetical scenarios in front of us.

Once we started focusing on what *caused* people to look at photos of a car, our research started producing many more fruitful insights. Now attuned to looking for the motivations that were driving behaviors, we quickly began seeing patterns.

CONTINUES ➤

CONTINUED ➤

One of the first patterns we noticed initially had us puzzled. As car shoppers would look through photos of a car, many would slow down to one photo—a close-up shot of the steering wheel of the car. They would spend several seconds intently studying the picture. Some would even zoom into the photo, or lean their head forward to try to get a better look at the steering wheel. When we asked people what they were doing, they replied that they wanted to see if the car was equipped with Bluetooth, a feature that enabled hands-free calling. To them, looking at a picture of the steering wheel was the best method to determine whether or not the car had Bluetooth, as a car with this feature would have a button to answer the phone on the steering wheel.

Our team was surprised by this, because the car detail page had a section devoted to listing the features and options the car came with. But as we learned from multiple interviews, car shoppers didn't want just a list of options and features; they needed to *see* the feature in order to feel confident that this might be a car worth purchasing. In other words, these car shoppers had a job in mind (find a car with Bluetooth), and to them, seeing a photo of the steering wheel was the best solution.

Once our team uncovered this job and the motivations behind it, we asked ourselves how we could design an experience that would better serve the job. Taking inspiration from a passage in Daniel Silverstein's *Innovator's Toolkit*, I created a one-sheet activity called the "Jobs to Be Done Canvas," shown in Figure 5.8. The canvas clearly defined the job at the top of the page and then split the requirements of the solution into two halves: the functional and the emotional.

Gathering the core team into the room, we filled the boxes based off what we had learned from customer interviews. At the top of the sheet, we defined the job: "When I'm shopping for a car, I want to definitively know the features it has."

In the "functional" box, we listed the tangible outcomes the solution needed to accomplish to satisfy the job. Team members would write ideas on sticky notes and add them to the canvas, like "show the feature, don't tell it" and "easy findability." On the other side of the canvas, we listed the "emotional" requirements, or how the customer should feel during and after the solution. During interviews, participants would tell us they wanted to feel absolutely certain that the feature was on the car, so we put things like "confident" and "assured" in the box.

By the end of this exercise, we had a single sheet we could turn to in order to understand the job and the requirements of the solution. This canvas became a living document that captured our findings in real time. After completing interviews with new participants, we would debrief in front of our JTBD canvases and make additions to the functional and emotional requirements based off of what we had just learned.

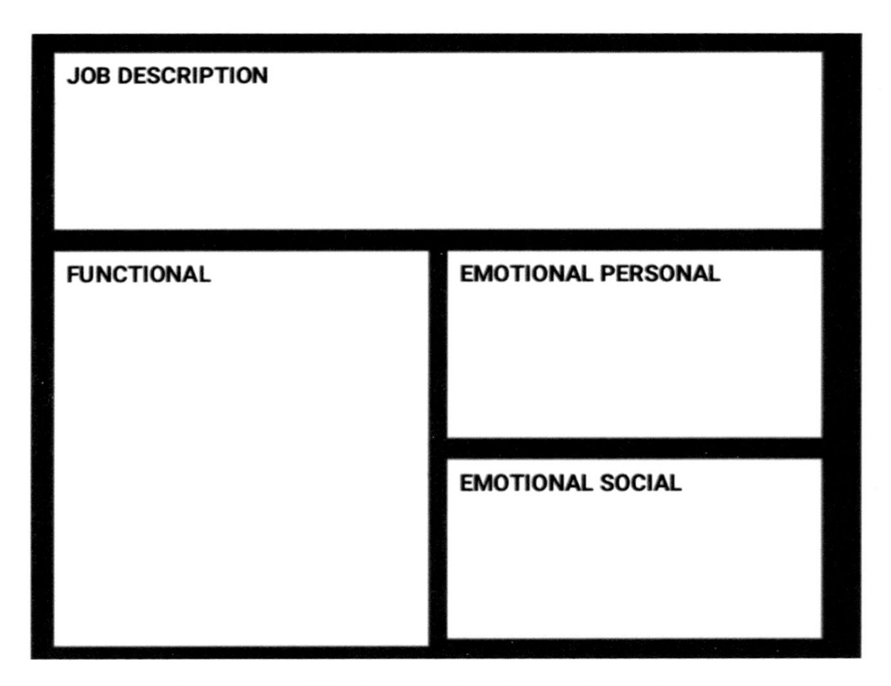

FIGURE 5.8 Designers and developers at CarMax used a simple canvas to understand the job to be done.

CONTINUES ➤

CONTINUED ➤

With this job understood, we launched a small change to the car page that produced a big result. Instead of providing a text list of the features on the car page, we displayed a series of thumbnail pictures of what the features looked like. This prevented people from having to hunt through all of the pictures of the car to find one particular feature, while also giving them the confidence that only a visual confirmation could provide.

By satisfying one of the jobs that car shoppers had when they visited carmax.com, we were able to serve their needs better, and thus make them more likely to purchase a car from us. Our business metrics increased when we split site traffic during an A/B test of features displayed in pictures versus the old way.

Energized by seeing the possibilities of designing for jobs, we continued using the framework and the JTBD canvas to uncover more customer insights. After launching the improvement in how we displayed car features, our team uncovered another job during customer interviews: understanding how big the car is.

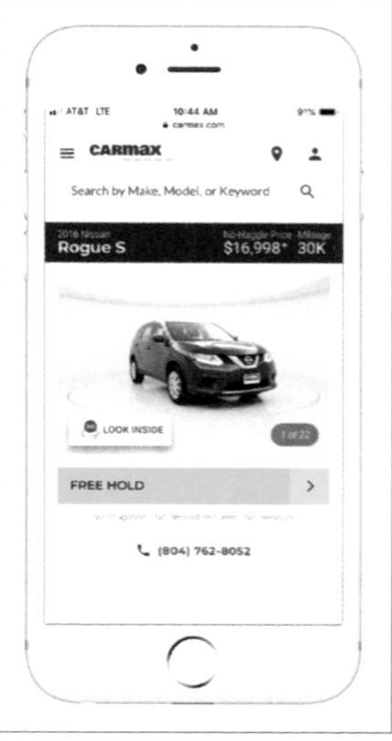

After looking through our analytics and noticing that a collection of interior shots of the front and back seat were highly engaged, we decided to dive deeper and understand why. We learned that these photos were valuable to shoppers trying to figure out how big the interior of the car was. Pictures, shoppers told us, gave them a better sense of the size and layout of the car better than other methods, like looking up the specifications or the cubic square feet. While this was a good solution compared to the alternatives, it still wasn't perfect.

Knowing now what we had learned from our first discovery of the unmet jobs on the car page, we recognized this as another opportunity to design for the job. We ideated our way to a new solution: a 360-degree photo of the inside of the car. This solution proved popular in testing, and the team quickly worked to implement it in all of our stores.

JTBD proved to be a radical mind-shift in how our team worked. We stopped trying to make shallow improvements to the interface, and instead went deeper to understand the motivations that drove shopper behavior. Thanks to JTBD, our team was able to focus on solving the biggest opportunities of the customer experience on carmax.com, thus making a meaningful impact to the product.

Jake Mitchell is a Principal Product Designer at CarMax, where he strives to reinvent the way customers find and fall in love with their next car. In addition to user experience design and research, Jake is proficient in web development and data science. This case study is a summary of his presentation "Using Jobs to Be Done at CarMax to Guide Product Innovation," given at UX STRAT 2017 in Boulder, Colorado.

Recap

JTBD not only helps you understand the customer's problem, but it also guides solution development. In particular, you can leverage JTBD in several ways to tie the design of products and services back to the individual's job to be done. This helps ensure that your solution is grounded in people's needs and will have a better chance of adoption.

JTBD helps guide the creation of a roadmap. *Product Roadmaps Relaunched* (O'Reilly, 2017) by C. Todd Lombardo, Bruce McCarthy, Evan Ryan, and Michael Connors shows how jobs can be a centerpiece of the process.

Job stories are short, encapsulated statements that reflect a job to be done with some added context. They guide the creative design of solutions using the job to be done as a key guiding light.

Just like buildings have blueprints, products and services have an underlying architecture, which isn't necessarily visible to the end user, but determines a lot of what they experience. JTBD guides the design of solution architecture by providing meaningful divisions and categories by which you can organize capabilities. In particular, User Environment Design (UED), as developed by Beyer and Holtzblatt, is an existing technique that resembles jobs thinking.

Finally, mitigate the risk of nonadoption by testing your assumptions. Lowdermilk and Rich show how you can use JTBD as a consistent element in your hypotheses to be tested. The key is to match jobs to the capabilities that help get that job done.

Ultimately, JTBD guides the transition from understanding your customers to coming up with solutions they really want. A simple focus on the job to be done drives core decisions from planning to design and development to testing assumptions.

CHAPTER

6

Delivering Value

IN THIS CHAPTER, YOU WILL LEARN ABOUT THESE PLAYS:

- How to create a customer journey map
- How to make customers successful during onboarding
- How to reduce churn
- How to provide better customer support

On December 3, 2001, the Segway was unveiled on the ABC News morning program *Good Morning America*. The first of the self-propelled transporters was delivered in early 2002.

The hype leading up to the launch of the Segway was immense. For instance, Dean Kamen, inventor of the Segway, expected to be selling 10,000 units a week by the end of 2002—that's half a million a year. Venture capitalist John Doerr also predicted it would reach $1 billion in sales faster than any company in history, and that the Segway could be bigger than the internet. Even Steve Jobs commented that the Segway would be as big a deal as the PC.

We were all supposed to be riding Segways by now. But by and large, the market rejected the invention. Although the transporter worked as envisioned, the Segway was doomed for many other reasons.

At $5,000 per unit, the Segway's price point targeted a more upscale market, excluding a large portion of potential buyers. Second, the Segway didn't fit into existing infrastructure. Do riders drive on the sidewalk while whizzing by pedestrians or join the traffic on the street at a much slower speed than cars? Police didn't know what to do with the Segway either. Legally, it wasn't clear where the Segway belonged.

Finally, and perhaps most importantly, there was a major social problem with the Segway: riders stood out in a crowd. They looked just plain silly—oddballs on a scooter. It was an awkward situation that invited mockery rather than amazement.

To be fair, the Segway has since found new uses. Tour groups ride them through cities to cover more ground while sightseeing. Park rangers rely on Segways to patrol their grounds while being more visible and accessible to visitors. Recently, smaller versions and even single-wheeled versions have made a comeback. But the fact is that the Segway did not transform the way we get around cities as predicted.

In the end, the market decides whether to adopt an innovation or not. The lesson here is that how an offering is introduced and presented to customers is just as important as getting the right solution. JTBD can help here, too—from understanding consumption jobs to reducing churn to providing better support.

PLAY ➤ Map the Consumption Journey

We live in a service-based economy, yet most organizations fail to provide good services. Good service delivery remains elusive, in part, because touchpoints with the customer happen over time and may be intangible. As a result, many teams turn to various mapping techniques to capture and diagnose opportunities in how they deliver solutions that customers value. In particular, journey maps show the interaction that customers have with your company and offering.

Journey maps are often conflated with job maps discussed in Chapter 3, "Discovering Value," but they represent different perspectives. A job map shows what job performers are trying to get done independent of a given solution. The aim is to understand their needs to increase your chances of adoption.

Instead of looking at a main job and its process, journey maps reflect what can be called *consumption jobs*, or the goals people have in finding a product or service, deciding to acquire it, and then getting value from it. Accordingly, buyers play a central role in mapping consumption jobs. These may or may not be the same as the job performer.

There are different ways to go about mapping the consumption journey. In many cases, a simple diagram of the steps taken to acquire a solution is enough. Figure 6.1 shows an example of a high-level consumption map created by Mike Boysen, a thought leader in JTBD.[1]

1. Mike Boysen, "A New Look at the Buyer Journey—as a Consumption Chain Job-to-Be-Done," *Medium* (blog), November 23, 2016, https://mikeboysen.com/a-new-look-at-the-buyer-journey-as-a-consumption-chain-job-to-be-done-fde28a6fa98d

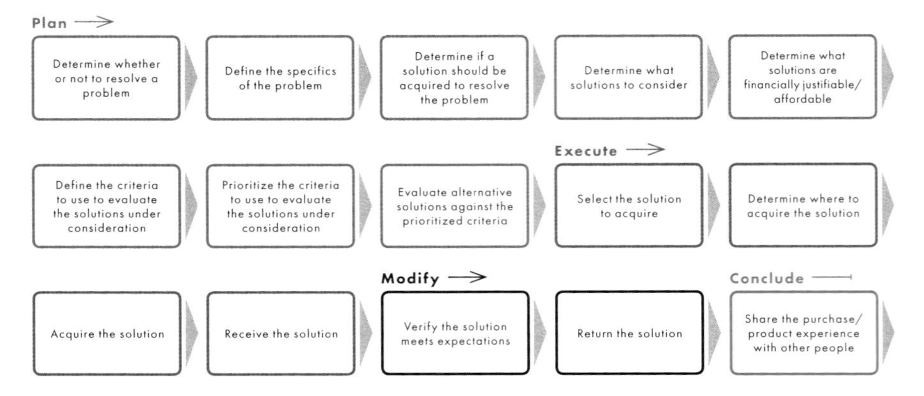

FIGURE 6.1 Simple consumption maps show the steps in finding and acquiring a solution.

Customer journey maps (CJMs) also visualize the consumption journey, but with more detail about other factors of the buying process, such as motivations, emotions, and specific interactions with a company or brand. Figure 6.2 shows an example of a typical journey map, in this case for purchasing health insurance, created by Jim Tincher, founder and CEO of Heart of the Customer, a consultancy specializing in journey mapping.[2] In the row labeled "Goals," you can see consumption jobs such as *learn about options*, *develop questions to ask*, *research plans*, *determine decision criteria*, *reduce choices*, and *make a final decision*.

CJMs usually include information about the touchpoints with a specific company or brand. For instance, in Figure 6.2 you can also see how a type of buyer navigates different channels to achieve their consumption jobs in the center portion of the diagram. This example also includes specific metrics and data about satisfaction and level of effort, providing a rich diagnosis of the consumption journey.

2. For more examples and resources on customer journey mapping, see Jim Tincher's company website: https://heartofthecustomer.com/

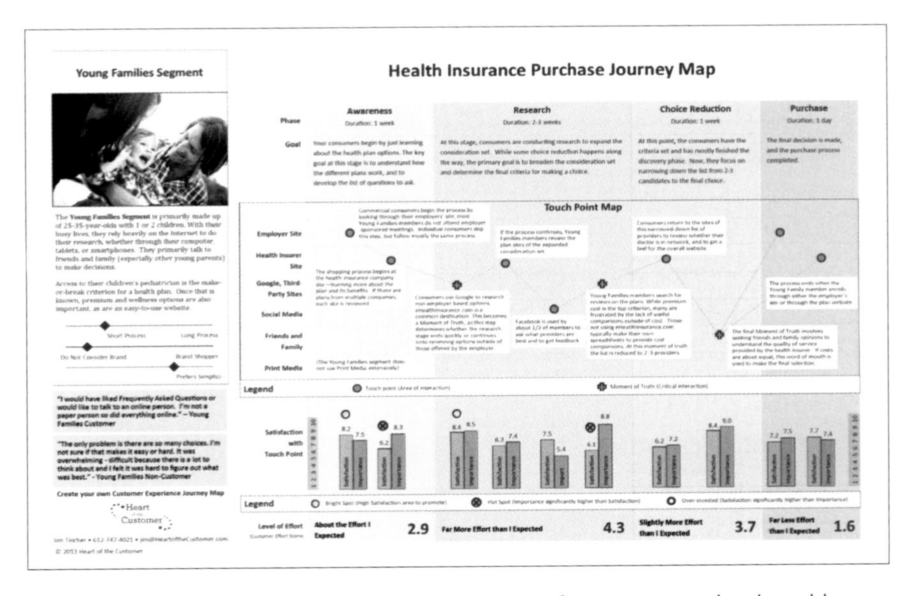

FIGURE 6.2 A journey map reflects the chronology of consumption jobs plus additional details of the buying process.

Journey Maps in Action

Regardless of your specific approach to mapping consumption jobs, the process can be broken into four separate steps.

STEP 1 ➤ Initiate a journey-mapping project.

Start by addressing three key questions:

- Whose journey are you mapping? By definition, a journey map looks at the buyer's experience. But you may also want to consider installers, technicians, and other decision-makers.
- What aspects of the journey are most useful to map? Start by identifying the consumption jobs that people have. But then consider other aspects, such as channels, emotions, satisfaction, and more if a broader picture of the buying context is needed.

- What are the bounds of the journey, i.e., when does experience begin and end? Strive to cover the entire consumption process, from realization of problem to solution selection to reasons for staying loyal.

STEP 2 ➤ Investigate the steps in consumption.

Ground the journey map in reality. Start by reviewing existing research, but then get out and talk directly to buyers. Interviews with six people are often all that is needed to flesh out an initial journey map. Larger samples are better.

Strive to uncover the sequence of consumption jobs, first to define the core flow of the journey map. Then research other aspects of the journey, such as emotions, pain points, and barriers to consumption. Consider other sources of data that inform the journey of your product, such as marketing metrics, survey data, and usage analytics.

STEP 3 ➤ Illustrate the journey in a diagram.

Distill the key elements from your investigation to complete the map. Focus on the individual's goals along the way. What triggers them to make a purchase? What do they want to optimize in their buying decision? What keeps them loyal?

Start by determining the main sequence of consumption jobs, relying on common phases:

- Plan
- Discover
- Learn
- Decide
- Purchase
- Setup
- Use
- Modify
- Upgrade
- Renew
- Leave
- Return

Longer, more descriptive labels are possible. It's important to phrase each one consistently, starting with a verb. Strive to make them as timeless as possible for longevity. For instance, instead of writing "See Fall 2018 TV Ad" under the "Become Aware" phase, use something like "View seasonal ads."

After you have a working model of the consumption jobs, add experiential details that reflect emotions and context. Create rows of information underneath these headers such as "thoughts," "emotions," and "pain points." Focus first on the objectives that consumers have along the journey: What are they trying to get done at each stage while interacting with your company? Also include other facets of information that will inform your team and business about the customer journey, including thoughts and feelings along this journey.

Your aim is to capture the as-is consumption journey first.

STEP 4 ➤ Align around the consumption journey.

Mapping the consumption journey is an important exercise in and of itself. But the real power of mapping is in fostering broader conversations within the organization around solutions and how to provide a better service. A journey map is a compelling artifact to engage others in dialogue (see Figure 6.3).

Plan different ways to make your journey map actionable. Schedule a workshop to interact with the diagram and then engage in creative problem-solving activities. Where are the biggest opportunities in supporting consumption jobs? How can you better support customers along the way? Where is the most friction in getting consumption jobs done?

After your team understands the current customer journey, determine how to make customers successful in getting their consumption jobs done. A separate to-be map may or may not be needed. Often, it's enough to map a future state alongside of the current state journey.

FIGURE 6.3 Mapping the consumption journey is a team activity, serving as a springboard into dialogue and collaboration.[3]

Overall, journey maps help instill a customer-centric mindset throughout your organization. They build a strong understanding of the customer's perspective, helping shift your view from inside-out to outside-in. They can reveal opportunities and risks in delivering value to your market. Job thinking helps guide the effort, with consumption jobs serving as the basis of the chronology. But rather than focusing on the main job and the individual's core objective, you'll shift your attention to how the person interacts with a brand or offering. This thinking may include consideration of emotional, social, and other jobs.

3. Photo of CareerFoundry offices (www.careerfoundry.com) used with permission from Martin Ramsin, co-founder and CEO.

LEARN MORE ABOUT THIS PLAY:
MAP THE CONSUMPTION JOURNEY

Lance Bettencourt, *Service Innovation* (New York: McGraw-Hill, 2010).

Bettencourt is a recognized leader in service design and innovation. In this book, the author provides a comprehensive set of tools and guidance for service innovation and delivery.

Tony Ulwick and Frank Grillo, "Can Bricks and Mortar Compete with On-Line Retailing?" (white paper, Harte Hanks, 2016).

This short document details an approach for mapping the consumption journey as recommended by Tony Ulwick and his company, Strategyn. Like his approach for mapping the main job, Ulwick looks at the steps involved in buying through a JTBD lens. The buyer's desired outcomes are then considered and prioritized at each step. The findings are based on in-depth qualitative interviews with buyers. Also, see Ulwick's full-length book on *Jobs to Be Done* (2016) for more on consumption job mapping.

Also see my 2016 book *Mapping Experiences* for more details and examples of mapping of all kinds.

PLAY ➤ Onboard Customers Successfully

In 1999, Salesforce reinvented sales and customer relationship management by providing a complete online solution. Previously, similar systems would have been installed locally, on servers at a customer's site. But with Salesforce.com, users could access data right from a browser with their data stored in the cloud.

Salesforce grew rapidly. In 2005, their market cap quadrupled from the previous year to $2 billion. By all measures, it was a wildly successful business. But they had a problem: as easily as customers could get started, so too could they leave. The company was losing 8% of its customer base each month—a nearly 100% loss in a year.

To address the situation, the company created a team that was labeled "customer success." Their mission: help customers get value out of the solution and keep them loyal. The results of their customer success team decreased cancellations significantly.

And so the field of customer success was born and adopted by other software as a service (SaaS) companies. The principle for subscription-based services was straightforward: help customers get their job done, and the business will grow.

Get Customers off to a Good Start

Onboarding customers into a subscription is critical to their long-term success. JTBD provides a consistent way to see how the objectives a customer is trying to achieve are independent of the solutions they use. Customer success managers ensure that customers get off to a good start, which leads to overall higher lifetime value.

For instance, Ryan Singer of Basecamp talks about how jobs thinking can help design better onboarding experiences. In an interview,[4] he points to two different perspectives of onboarding:

> ...[Distinguish between] "helping people be better at using your product" versus onboarding as "helping people be better at what your product lets people do." The more you have to explain the product, the less attention you can have on why people use it and, more importantly, what they have to do differently... Proficiency with software is never the goal, anyway. Instead, it's something external; it's the job people are trying to get done.

In other words, onboard customers into the job, not only into the service.

4. See Samuel Hulick, "Applying Jobs-to-Be-Done to User Onboarding, with Ryan Singer!" *UserOnboard* (blog), 2016.

Alan Klement echoes Singer with a concrete framework to identify and design for different onboarding scenarios. In his article, "Design for Switching: Create Better Onboarding Experiences," Klement shows that to optimize your onboarding, think of the two separately:[5]

- The **solution experience** dimension (shown horizontally in Figure 6.4) considers how familiar customers are with your product or service. It considers the skills needed to use a tool.
- **Job comprehension** (shown vertically in Figure 6.4) reflects how well customers are knowledgeable and skilled in getting the job done.

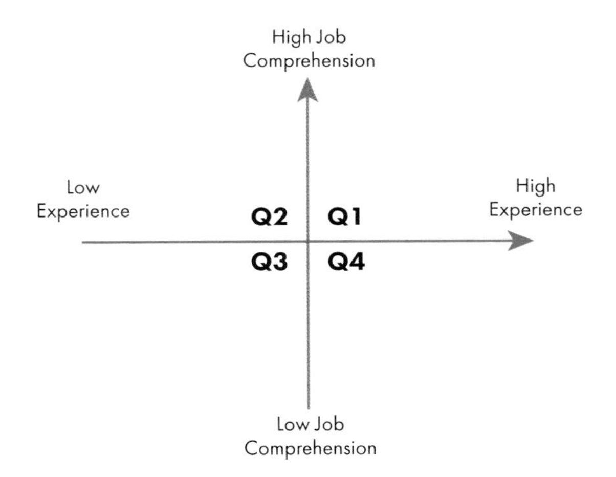

- **Solution Experience:**
  How familiar they are at using solutions (products) to solve a job
- **Job Comprehension:**
  How well a customer understands the Job to Be Done

FIGURE 6.4 An assessment matrix for different types of users during onboarding.

5. Alan Klement, "Design for Switching: Create Better Onboarding Experiences," *JTBD News* (July 2014).

Klement's typology yields four different onboarding scenarios:

- **High-solution experience/high-job comprehension (Q1):** This group has a thorough grasp of the job, as well as other solutions. In this case, your intent is to demonstrate value beyond your solution and educate users on the deeper aspects of getting the job done.
- **Low-solution experience/high-job comprehension (Q2):** In this case, users have filled out expense reports using a different solution. They know why they are important and how to complete the job. You'll want to focus on explaining the tool to them and, in particular, the differences your solution has to others.
- **Low-solution experience/low-job comprehension (Q3):** The onboarding process needs to be gentle, starting with an overall description of why getting the job done is important. You may need to walk users through the entire process step by step, explaining the features and functions as you go.
- **High-solution experience/low-job comprehension (Q4):** It's possible but less common that users know a lot about a solution and little about the job to be done. Still, in this case, you can assume that people have a limited knowledge of the tool itself and only use part of its capabilities. People become more sophisticated in what they expect, and therefore perceive to be getting more value from your product.

Knowing which category users fall into allows you to target messages that address a given user's gaps in knowledge for a better initial experience. Here's how it works:

STEP 1 ➤ Learn about the customer.

Assess new users as they initially come into contact with your solution. Either through a survey or directly in person, ask them what their experience levels are for both the tool and getting the job done. For example, if you are providing an online solution for employees of a company to submit expense reports, you could ask them a few simple questions while signing up for the service. Figure 6.5 shows the type of information you can collect.

Have you ever created expense reports before?

yes ●——— **no**

Have you used them to...

- ☐ Manage different expense accounts
- ☒ Manage a big business
- ☐ Get approval for logged expenses
- ☐ Upload images of receipts

Questions that help determine Job Comprehension

FIGURE 6.5 Example questions to ask to assess job comprehension and tool comprehension.

Based on the responses, determine which quadrant of the matrix in Figure 6.4 each user falls into.

STEP 2 ➤ Determine the optimal sequence of tasks for each learning type.

Next, sequence the steps you'd like to have customers take, not only to use your solution, but also to reach their goals. Strive to reduce the time to value. Focus on what customers struggle with most and what gaps in knowledge typically exist. Recognize that people will need different types of information to guide them into your solution.

Your goal is to get customers to complete a task that is both relevant to learning how to operate the solution and getting the job done. If you can combine these, you kill two birds with one stone. For instance, at MURAL, we demonstrate how to use our software (solution experience) by conducting a short remote brainstorming session (job comprehension).

STEP 3 ➤ Design the onboarding experience.
Create content and messages that target each of the different scenarios. Develop features and functions to support the different scenarios.

Instead of just telling people what actions to take, give clues about the outcomes they can expect. The words and phrases you choose need to reassure them that your solution will help them get their job done.

Consider the examples in Table 6.1, which takes tool-centric statements and makes them jobs-centric statements. Include information about the job that will be completed and the benefit that users will see. What are your customers' desired outcomes? How will they measure success?

TABLE 6.1 EXAMPLES OF JOB-CENTRIC MESSAGES DURING ONBOARDING

SOLUTION-CENTERED MESSAGE	JOB-CENTERED MESSAGE
Automate filling out expense forms with our auto-scan recognition software.	Save time submitting expense reports using auto-scan.
Click on "Total" to add up your expenses with different exchange rates.	Improve the accuracy of your expense reports with automatic updates of the most current exchange rates.
Select "Submit" to send your expense report for approval.	Ensure that the right person has received your expense report any time you click "Submit."

Focusing on underserved needs, in particular, will increase the chances of messages resonating with customers. What are the most critical desired outcomes that people have? Highlight those needs in your explanation of the solution during onboarding.

Extending the Technique

Onboarding doesn't necessarily start after customers subscribe to a service. You can also consider touchpoints while they learn about and consider your solution as part of a type of pre-onboarding. In these situations, you can also focus on the job to be done to go beyond the post-purchase experience.

Consider the following two descriptions of the same camera. The first one is from Amazon, shown in Figure 6.6. It details the specifications of the camera, including things like "SnapBridge Bluetooth Connectivity," "24.2MP DX-Format CMOS Sensor," and more.

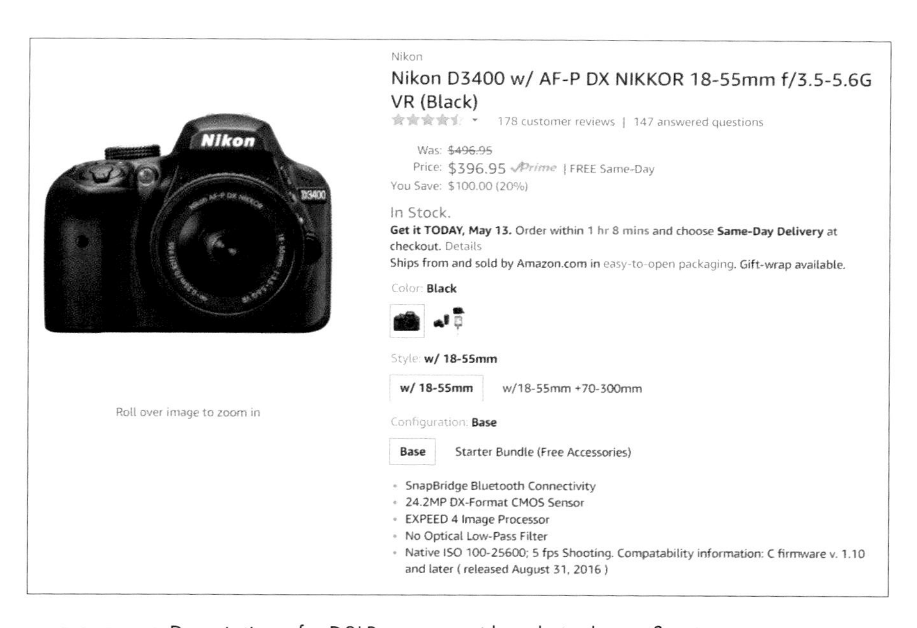

FIGURE 6.6 Description of a DSLR camera with technical specifications.

These technical details are important to some buyers, particularly serious photographers who need to know this type of information. But what is the job to be done? What do people use sophisticated DSLR cameras for these days and what is the process of doing it?

Figure 6.7 shows the same camera on the Best Buy website. It also includes technical specifications, but, in addition, speaks to the job to be done. The description below the camera image in the center reads: "Share images straight from your DSLR with the Nikon D3400. Instant connectivity to your smartphone lets you share photos to your favorite social media platform without having to physically change cards."

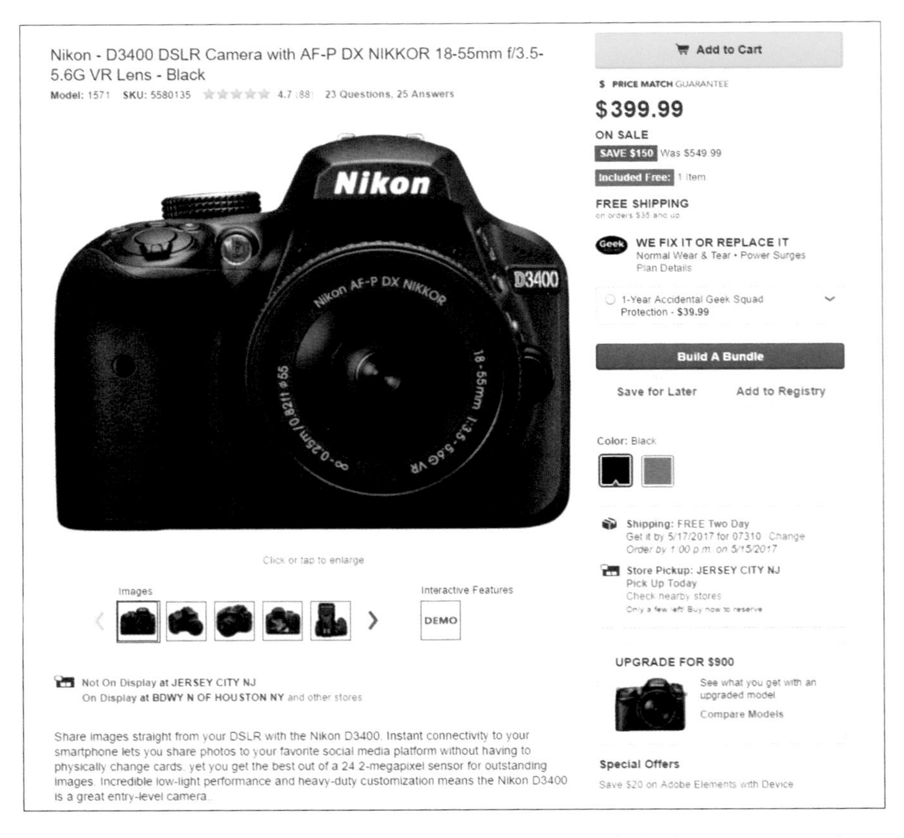

FIGURE 6.7 Description of a Nikon camera on Best Buy includes language around the job to be done.

LEARN MORE ABOUT THIS PLAY:
ONBOARD CUSTOMERS SUCCESSFULLY

Alan Klement, "Design for Switching: Create Better Onboarding Experiences," *JTBD News* (July 2014).

> In this detailed post, Alan Klement lays out a typology of onboarding experiences based on JTBD. Two dimensions—job comprehension and solution expertise—form a simple matrix with which to segment users. Then, applying Switch interview techniques, you can uncover where users fall on the matrix. While logical and practical, Klement does not offer any evidence or case stories of how his technique has been used successfully.

Samuel Hulick, "Applying Jobs-to-Be-Done to User Onboarding, with Ryan Singer," UserOnboard (blog), 2016.

> Singer discusses how JTBD can help think through onboarding design in this interview, using Basecamp—the product he manages—as an example. There isn't much practical information in this piece, but Singer makes an important distinction between onboarding into the job versus onboarding into the tool. This is a short read, and very much worth checking out.

PLAY ➤ Maximize Customer Retention

Consider for a moment all of the services you can subscribe to these days: music, file storage, rental cars, coffee, socks, and razors, among many other things. Even businesses have subscriptions, ranging from CRM systems and analytics to employee payments and HR functions. The list goes on.

Beyond giving customers a great onboarding experience, the key to success in the subscription economy is to help customers get their jobs done over time. Lincoln Murphy, a thought leader in customer success, defines the field as "when your customers achieve their Desired Outcome through their interactions with your company."[6] In other words, address the needs that people have after "hiring" your services, and both the customer and the company will benefit.

The impact on businesses is profound. Subscription-based businesses now have an imperative to not only sell more and more to customers, but also build long-term customer relationships. After all, the competition is only one-click away, and customers can always cancel subscriptions.

Cancellation Interviews

The cost of acquiring a new customer typically outweighs the cost of keeping existing customers by a factor of five or more. Preventing customers from switching increases their overall lifetime value. When customers cancel, they are said to *churn*. Businesses need to understand and manage churn, keeping it to a minimum.

Typically, companies will ask departing customers to select a reason for leaving on a simple form. The results often reveal such reasons as the following:

- Not using the product anymore
- My project was over
- No need

While such information provides some basic clues into the causes of churn, it doesn't go nearly deep enough. In most cases, a more thorough understanding is needed to be able to take action to prevent churn. Exit interviews can be helpful to uncover hidden motivations. Here, too, jobs thinking can help you get to the core of the issue.

Specifically, you can use Switch interviews, described in detail in Chapter 3, to probe into the underlying reasons why subscribers cancel.

6. See Lincoln Murphy, "Understanding Your Customer's Desired Outcome," *customer-centric growth by lincoln murphy* (blog), no date.

However, instead of asking new customers why they switched *to* your solution, you should ask why they switched *away* from your service. Think of this technique as an inverted Switch interview.

Often, you'll find that the decision to leave didn't take place the day of cancellation. Instead, there were a series of events and decisions that happened weeks, months, or even years before that led up to the cancellation. Dig deep to recreate that decision-making process and find the root cause. Here's how to learn more about reducing churn and avoiding losing customers using JTBD.

STEP 1 ➤ Conduct cancellation interviews.

Start by identifying and contacting customers who have churned. Using a timeline such as the one below, hold an open discussion to work backward to the first thought—when did the customer first consider cancelling?

- **Consumption:** When and how they consumed the service when they were using it.
- **Cancel Moment:** This occurs when they evaluate options and actually buy/cancel. Pay attention to trade-offs.
- **Event Two (Active Looking):** Usually, this is a time-sensitive event that triggers active looking.
- **Event One (Passive Looking):** Something happens here that gets them to start considering alternatives, although probably not actively looking.
- **First Thought:** This is when they first had the sense that your solution wasn't fulfilling their needs. It's the earliest moment in failing to get a job done.

Ask probing questions to your former customers, such as the following:

- What prevented you from getting your job done?
- What desired outcomes remained unmet?
- When did you start to have problems or doubts with the service?
- Why did you start to evaluate something else?

Focus on the barriers they had in getting their job done. Often, it's not the same reason that they gave on your cancellation form. You'll likely learn a great deal more about their motivations for leaving, which may not be related to just your offering. Also understand the broader circumstances of getting the job done.

STEP 2 ➤ Find patterns in cancellation reasons.

Your overall aim is to find patterns that you can leverage to make improvements. It's not about trying to win back customers one-by-one, but to consistently prevent churn from happening in the first place. Think of this approach to interviewing as a type of root cause analysis across customers.

Here are some questions to reflect on with your team about the insights in the interview data:

- At what point do people decide to abandon your solution?
- What are common barriers you could change to avoid cancellation?
- What part of their job process is most problematic?
- What needs do they have that are not getting addressed?
- What circumstances influence customers who are leaving?

STEP 3 ➤ Address the root causes.

Typically, issues upstream—during onboarding, purchasing, or even earlier—have an effect later in the customer lifecycle that lead to churn. Once you've found patterns, ideate solutions to resolve the root causes. Take the issues you identified and turn them into "How might we…?" statements. Come together as a team to find solutions to churn triggers.

For instance, Ruben Gamez, founder of Bidsketch, an online tool for creating proposals, did cancellation research for his company's software. By reversing the Switch interview technique, he was able to trace the first thought that customers had about cancelling. Gamez wrote about several patterns, and the team then devised ways to address them, including the following:[7]

- The team believed that showing custom template designs during onboarding would help expose the solution's value to customers better.
- The primary emotional job that was unfulfilled was giving customers confidence that their proposals were well-designed. Accordingly, they came up with ways to provide that confidence at each step of using the product.
- Reducing the size of a subscription was a clear sign that customers had already decided to cancel. Proactive outreach at the first thought of downgrading could prevent churn later.
- The company was eventually able to recognize early warning signs better and take action to address possible churn with education and resources.

In the end, people stay loyal to reaching their objectives. Of course, they need to be able to use your solution, but that's just the first step. Your retention strategy needs to also consider the work that customers are trying to complete and how to make them successful. Only then can you consistently prevent churn and increase customer retention.

Extending Understanding of Retention

The Four Forces technique outlined in Chapter 3 also provides insight into retention. In particular, the factors on the left side of the diagram (see Figure 6.8) show the key reasons why people stay.

7. Ruben Gamez, "Doing SaaS Cancellation Interviews (the Jobs-to-Be-Done Way)," *ExtendsLogic* (blog), 2015, www.extendslogic.com/business/jobs-to-be-done-cancel-interviews/

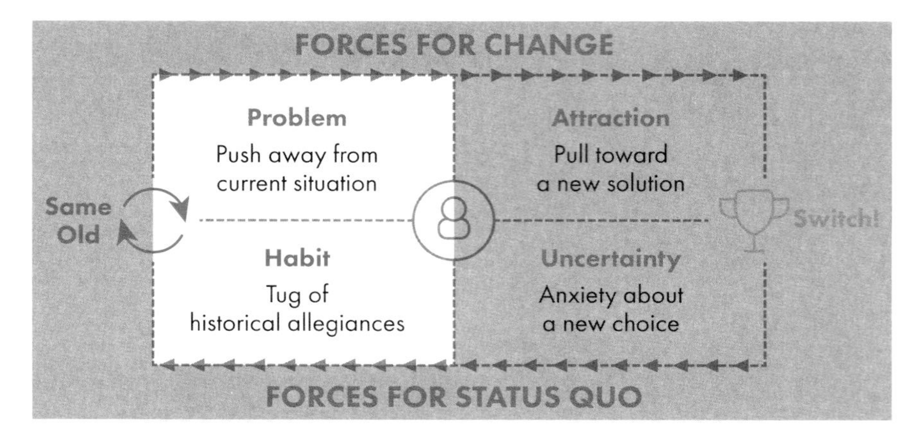

FIGURE 6.8 Retention can be viewed as a combination of reducing problems pushing customers away and increasing the habit-forming aspects of your solution.

On the one side, you want to reduce factors that may push customers away from your solution. You should constantly strive to reduce the customers' friction of interacting with your product and your company. Listen carefully for aspects of the service that may cause pain and detract from the benefit that attracted people to your offering in the first place.

On the other side, find ways to increase your customers' sense of familiarity with the service. The goal is to promote regular use of your service by providing habit-forming features and exercises. But make no mistake: changing an existing behavior is extremely challenging to control.

B. J. Fogg, psychologist and author from Stanford University, provides some of the best guidance and understanding of behavior change needed to create a habit. The Fogg Behavior Model (Figure 6.9) shows that three factors must be true for a new behavior to occur:

- **Motivation:** In terms of JTBD, a key motivation is the desire to accomplish an objective. Understanding the job gives great insight into motivation. When motivation is high, people are more willing to change.

- **Ability:** People must have the skill to perform the job to be done. Your solution should augment their natural abilities and teach them how to reach their goals.
- **Prompts:** A trigger is needed for the target behavior to occur and the habit to form. Giving your customers a clear call to action is crucial in creating a new habit. In most cases, taking baby steps and slowly revealing steps of the new behavior work most effectively.

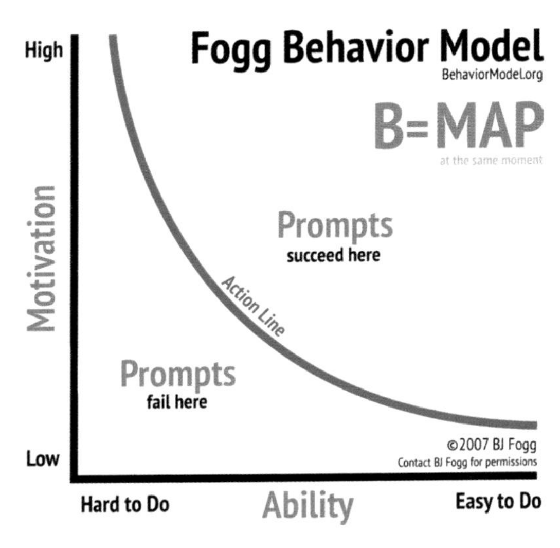

FIGURE 6.9 The Fogg Behavior Model shows how motivation, ability, and prompts come together to affect a behavior change.

When a behavior change does not occur, at least one of those three elements is missing. The aim in increasing retention rates is to understand the factors that drive habits and build them into your offering.

For instance, imagine you are the provider of online task management software. The value of the software goes up if people use it on a regular, daily basis. To *motivate* and inspire them, you may share case studies and examples of how other teams have benefited. You can also highlight milestones of use within the product to motivate as well, such as including badges or small rewards for completing tasks. To increase users' *abilities*, you might want to consider online courses, both live and self-paced. But you can also include in-product tutorials and learning devices to help them understand the workflow. Finally, *prompts* might come in the form of notifications into other streams of communication, such as email or group chat. In the end, it's a combination of all three that leads to habit-forming behavior.

LEARN MORE ABOUT THIS PLAY:
MAXIMIZE CUSTOMER RETENTION

Ruben Gamez, "Doing SaaS Cancellation Interviews (the Jobs-to-be-Done Way)," *ExtendsLogic* (blog), 2015, www.extendslogic.com/business/jobs-to-be-done-cancel-interviews/.

This post is packed with practical information about running cancellation interviews based on the Switch technique. It includes details about recruiting, running interviews, and how to follow up. Ruben also provides real-world examples of using cancellation interviews in practice. Also see Chapter 3 for more on Switch interviews and references there.

B. J. Fogg, "Fogg Behavior Model," 2019, https://www.behaviormodel.org.

This website is the official resource about Fogg's behavior model with links to further materials and tools. Fogg founded the Behavior Design Lab at Stanford University and regularly teaches industry innovators how to change behavior. See also his Tiny Habits website and course, designed to create new habits in small steps: https://www.tinyhabits.com/join

Brian Rhea, "Customer Acquisition & Customer Retention" (blog), March 5, 2019, https://brianrhea.com/customer-acquisition-customer-retention/.

This article outlines how the Four Forces technique can provide insight into both customer acquisition and retention. The author offers a short, but very helpful video to explain the concept. Rhea is an active thought leader and practitioner of JTBD techniques with a wealth of useful advice on his website.

ADOPTION OF INNOVATION

Building a better mousetrap does not guarantee that people will adopt an innovation. Many would-be breakthroughs fail for not taking human factors into consideration. You don't need to look further than the nonadoption of offerings like the Segway, Zune, Amazon's Fire phone, and even Google Glass. To increase chances of adoption, organizations must understand people's needs and desires.

Everett Rogers pioneered research into how innovations get adopted. His landmark book *Diffusion of Innovations* is the culmination of decades of rigorous research.[8] In it, he describes five key heuristics that predict the adoption rate of innovations:

- **Relative advantage:** Is the innovation better than existing alternatives to getting a job done?
- **Compatibility:** Is the innovation appropriate? Does it fit into the user's daily life, beliefs, and values?
- **Complexity:** Is the solution easy to comprehend and use?
- **Trialability:** Can it be tested without penalty?
- **Observability:** Can the innovation be observed and understood?

The above factors describe how an adopting population is likely to perceive the change brought on by introducing an innovation. If an innovation is too complex to use and hard to understand, for instance, it may not get adopted. Or, if an innovation is contrary to one's beliefs (i.e., not compatible), it may also get rejected.

In the case of the Segway, for instance, there were major compatibility issues with existing roadways and existing laws. Observability was also high, but that visibility backfired and brought ridicule rather than reward. Some simple market tests might have identified these barriers to adoption.

In the end, innovations that get a job done for the end consumer have a higher chance of succeeding. Focusing on the job to be done isn't just about creating better products— it's about whether your business will ultimately survive or not.

8. Everett Rogers, *Diffusion of Innovations*, 5th ed. (New York: Simon & Schuster, 2003).

PLAY ➤ Provide Relevant Support

"Get me to my mother's house!" That's what I demanded over the phone to an agent at Zipcar, a leading car-sharing provider, on Mother's Day one year. He and I seemed to have a misunderstanding about what I wanted to get done.

The conflict started when I arrived at the spot of my reserved car, and there was no car there. After failing to locate the vehicle I booked, the agent proceeded to name alternate locations where I could get a different rental car. All were too far away and would have added hours to my journey, making me late.

The pain in this interaction came from a mismatch in goals, I believe. I was trying to get to my family's house at a specific time on Mother's Day. He was trying to rent me another car, even if that didn't meet my objective. As a result, our exchange was longer than needed and was more contentious. We weren't seeing eye-to-eye.

Eventually, the agent was able to find another car, but not for the same time period. We arrived late and had to leave early. Zipcar failed to help me accomplish my job to be done that day.

My interaction with Zipcar begs the question: What business are they really in? Does Zipcar just rent cars, or do they provide reliable mobility? How might the agent have helped me differently? And how might everyone in the company align around the customer need in order to serve customers and ultimately to innovate?

JTBD provides a common point of view for organizations to understand customer goals. As we've seen so far in this book, knowing these objectives can help you define and design offerings customers find valuable for the market. But jobs thinking can also be pervasive throughout an organization, filtering all the way down to support agents. The results of applying jobs thinking can help even simple interactions like I had with Zipcar. In other words, everyone in an organization can be aligned to customer needs through JTBD.

Jobs-Driven Customer Support

In support situations, people don't necessarily ask for what they want directly. For one, they may use the wrong language. How should customers know the jargon and terms associated with your solution? Additionally, they may also have already devised a solution to their problem in their mind and ask about getting the wrong job done. Agents have to be careful and clarify the real need before trying to come up with a solution.

Active listening is a big part of providing good support. But beyond that, it's also about understanding the customer's intent and resolving the core issue.

STEP 1 ➤ Listen for the job.

Support agents need to listen carefully. Remember that customers may use different language or even confuse language, labeling things the opposite way that you might. Their goal: uncover the JTBD. But people often focus on the task at hand and the technology. To do that, you have to hear what's behind the words.

STEP 2 ➤ Clarify and assess.

Determine the level of detail of their problem and clarify the situation. Ask "how" to get more specific and "why" to get more general. Typically, support questions are specific, and you'll have to ask "why" to uncover the underlying job. In other cases, customers may give you a broad issue that you need more detail on to solve. Ask "how" to drill down further.

Regardless of the solution, reflect back to the customer what you believe their JTBD is. They need to know that they've been heard. Confirm that you're on the same page. For instance, during my interaction with Zipcar described earlier in this section, it would have been nice to have heard, "I understand you need to be somewhere on time, and I'm trying to help you." But my real expectation was that he'd offer to pay for a cab or Uber to my destination. (They'd done that in the past.) Instead, this time the agent was intent on finding a replacement vehicle nearby even if that didn't get my job done.

STEP 3 ➤ **Resolve the issue.**

Consider what you could do within the scope of your abilities to address the JTBD. This may not be exactly what the customer asked for initially. They may have already come up with a solution to their problem before even contacting your organization. But if their knowledge is limited, they may be asking for the wrong thing. Additionally, they may be using the wrong thing to make their inquiry.

By taking a step back and considering what they are trying to accomplish, your support agents can often resolve issues in a more satisfying manner. Note that often you'll be exploring steps in the job process or even smaller micro-jobs. Still, jobs thinking lets you view their objective in the moment, independently of your product, before finding a solution to their request.

For instance, let's say you work for a provider of online project management software and a customer calls about difficulties with the download function. After a few short questions, you then discover that the customer would like to print out a document to be able to display the content as a poster in a workshop. While resolving the download issue for the customer is worthwhile, you might also point out that your software has full-screen mode to project documents, as well as the ability to invite others as guests to the online version of it.

By realizing that the customer wants to *share content with colleagues during workshops*, you can better resolve their core issue. Taking a step back and considering the job to be done opens up new solutions and potentially improves the customer's future use of your product.

JTBD for Online Event Organizers

Kathryn Papadopoulos, Research Manage at Google, and Jim Kalbach, author and Head of Customer Experience at MURAL

Online events are a popular way for brands to engage their markets. These days, companies can reach more and more customers with webinars, virtual conferences, and online meet-ups. To understand the job of running online events, my design team at GoToWebinar, a leading webinar platform, conducted jobs to be done research.

At the time, I was the design practice representative for the offering steering committee, which consisted of an offering director, product manager, engineering lead, and marketing lead. I worked together with our lead user researcher, Kathryn Papadopoulos. We wanted to understand customer jobs to be done to identify opportunities for innovation, but also how to address the market and deliver value. Our process had four overall phases.

1. DISCOVER JOBS.

After defining the main job performer, *online event organizers*, and the main job, which was to *conduct an event online*, we began with in-depth interviews. We spoke with eight organizers of online events and discussed their process. Each interview was recorded and then transcribed.

From the qualitative data gathered, the team created a mental model diagram instead of a chronological job map, shown in Figure 6.10. This began with planning steps, such as "Gather requirements" and "Reach audience," before moving into jobs around the event itself, including "Start event" and "Engage attendees." The diagram ended with concluding steps around following up and improve events for next time.

CONTINUES ➤

CONTINUED ➤

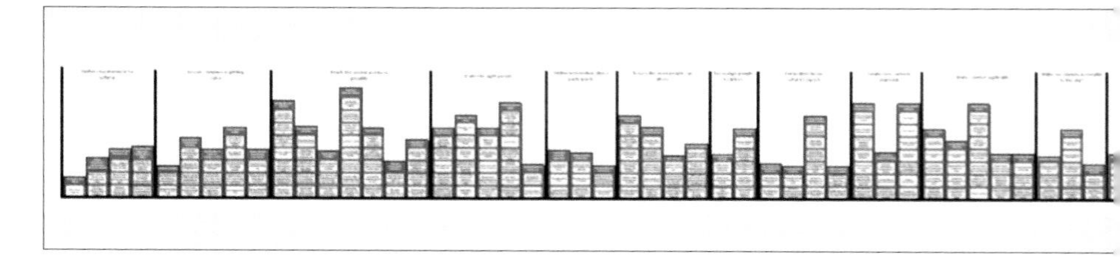

FIGURE 6.10 The top half of a mental model diagram for running online events.

2. VALIDATE THE NEED STATEMENTS.

From the data and the diagram, we also generated about 50 need statements. We refined these, and then in a second round of interviews with a total of six participants, we validated the outcome statements with job performers. Each interview lasted about two hours.

Kathryn devised a clever way of discussing their process without reading off the statements. If the participant mentioned the theme of a need, we recorded it in a spreadsheet. This allowed us to review the overall frequency of mentions to gauge validity. Note that in some cases, participants mentioned the opposite of our recorded need, which we then discussed how to deal with, primarily through qualifying or reformulating the original statement (see Figure 6.11).

After getting feedback from job performers, we reformulated a few of the need statements, removed one or two, and split the others. All in all, the changes to our original set of needs were minimal.

3. PRIORITIZE NEEDS.

Next, we prioritized the need statement with our customers in a survey. This followed the ODI method, as developed by Tony Ulwick and colleagues at Strategyn. (See the Play,"Find Underserved Needs," on p. 84.)

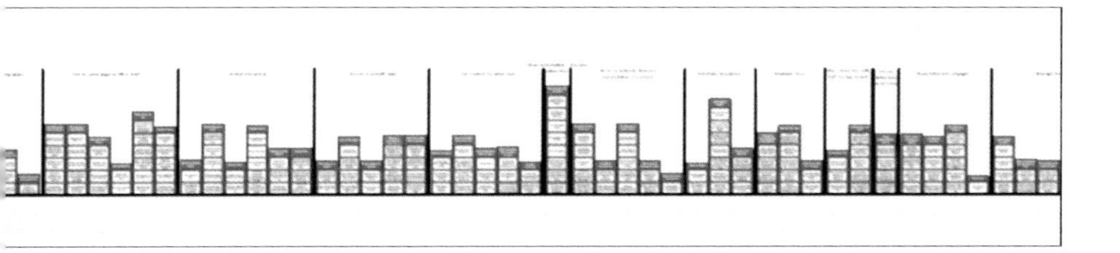

Each need statement was paired with a set of questions:

1. How important is this outcome to you, on a scale of 1–10?
2. How satisfied are you with this outcome, on a scale of 1–10?

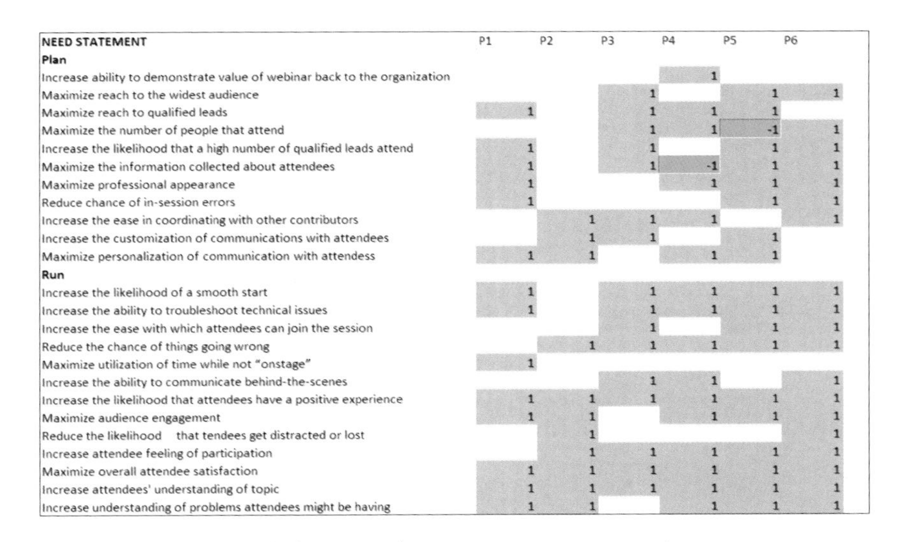

NEED STATEMENT	P1	P2	P3	P4	P5	P6
Plan						
Increase ability to demonstrate value of webinar back to the organization				1		
Maximize reach to the widest audience			1		1	1
Maximize reach to qualified leads	1		1	1	1	
Maximize the number of people that attend			1	1	-1	1
Increase the likelihood that a high number of qualified leads attend	1		1		1	1
Maximize the information collected about attendees	1		1	-1	1	1
Maximize professional appearance	1			1	1	1
Reduce chance of in-session errors	1				1	1
Increase the ease in coordinating with other contributors		1	1	1		1
Increase the customization of communications with attendees		1	1		1	
Maximize personalization of communication with attendess	1	1		1	1	
Run						
Increase the likelihood of a smooth start	1		1	1	1	1
Increase the ability to troubleshoot technical issues	1		1	1	1	1
Increase the ease with which attendees can join the session			1		1	1
Reduce the chance of things going wrong		1	1	1	1	1
Maximize utilization of time while not "onstage"	1					
Increase the ability to communicate behind-the-scenes			1	1		1
Increase the likelihood that attendees have a positive experience	1	1	1	1	1	1
Maximize audience engagement	1	1		1	1	1
Reduce the likelihood that tendees get distracted or lost		1				1
Increase attendee feeling of participation		1	1	1	1	1
Maximize overall attendee satisfaction	1	1	1	1	1	1
Increase attendees' understanding of topic	1	1	1	1	1	1
Increase understanding of problems attendees might be having	1	1		1	1	1

FIGURE 6.11 A record of mentions by six participants in a needs validation study.

CONTINUES ➤

CONTINUED ➤

From this, we were able to compute opportunity scores, which pinpointed needs that were important but had low satisfaction.

In the end, we had a list of needs that represented unmet needs, as follows:

- Reduce the chance of in-session errors.
- Increase the likelihood of a smooth start.
- Increase the ease with which attendees can join the session.
- Increase the ability to troubleshoot technical issues.
- Maximize the number of people who attend.

Before administering the survey to job performers, we also surveyed the GoToWebinar leadership. What needs did they think were most important to customers? This allowed us to compare which need they thought provided the most opportunity compared to customers. Only one team member correctly guessed two of the top five needs prior to the survey; everyone else correctly picked one of the top five unmet needs.

We further segmented participants by type. We found different purposes for running online events. Some involved conducting lectures or information-sharing sessions. Others were giving trainings. The largest segment was event organizers holding online sessions for marketing purposes.

4. TURN INSIGHT INTO ACTION.

Overall, the research helped the steering committee in several ways. First, we contributed to a pricing and packaging effort already underway. We were able to map features that made up the different packages (e.g., custom registration) and informed the different pricing levels.

Second, we were able to recalibrate marketing messages to reflect customer jobs. This shifted our language from "what" to "why." For instance, instead of "Enjoy our one-click registration," we suggested "Allow attendees to join easily with one-click entry" to reflect the desired outcome.

Finally, the insights from our JTBD research also helped prioritize the product development roadmap. Our research confirmed some of the work that was already planned, but also brought forward development we'd previously overlooked.

One offering leader said, "The JTBD research and findings were a big help in shaping the offering. It gave us confidence that we were relevant to our customers' needs and provided concrete guidance.

Kathryn Papadopoulos is currently a UX Research Manager at Google, working across Search, Assistant, and News product areas. She focuses on evolving UX research tools and processes that enable inclusive design and quick iteration, like the Google Research Van (a mobile lab on wheels) and Rapid Research models (one-week turnaround projects). Prior to Google, she worked as a researcher at Citrix and was a program manager in the Makerspace and Online Education fields at Stanford.

Recap

Understanding the customer, not the competition, is the key to delivering offerings that customers value. JTBD approaches can be applied in various activities in a go-to-market strategy.

A journey map illustrates how customers will consume an offering. This differs from a job map in that it is viewing customers in relationship to a company or brand. People involved in sales and marketing can benefit from understanding the customer's journey, as well as product design and development and other disciplines.

Once customers decide to purchase your solution, you may need to onboard them into the product or service. JTBD can help guide the steps taken to make customers successful in getting their job done.

It's generally much less costly to keep an existing customer than to acquire a new one. Retaining customers requires that they adopt your innovation, get value, and ultimately become advocates for you. Even the best inventions fail because they don't get adopted. Customer success is a practice that looks at helping customers not only use a solution, but also ultimately get their job done.

JTBD insight and techniques can cascade down through an organization. Support agents, for instance, can resolve customer issues better by focusing on the job to be done while interacting with customers. Even the structure of help desk articles can be influenced by the elements of JTBD.

CHAPTER

7

(Re)Developing Value

IN THIS CHAPTER, YOU WILL LEARN ABOUT THESE PLAYS:

- Ways to survive disruption using JTBD
- How to create a jobs-based strategy
- How to organize around JTBD
- How to expand your business through aspirations

When Skype first launched in 2003, it had limited functions, and the quality of the calls was poor, but it was free. Users just had to sign up, and they could speak with each other around the world.

Incumbent web conferencing providers, such as WebEx and GoToMeeting, largely ignored the service. "We're suited for business customers. Skype is for college kids to chat with one another," they'd say or something similar. Established telecommunication companies thought nothing of Skype either. After all, it was a free service that targeted the lower end of the market. Why should the big players care?

Then in 2011, Microsoft acquired Skype for $8.5 billion. Skype's market share of all international calls nearly tripled that next year. And a few years later, Microsoft launched Skype for Business. Now, Skype was on equal footing with other premium services.

Even telephone providers were feeling the effect, with the per-minute cost of international calls being commoditized to near zero. Skype also changed the playing field in terms of security, technical architecture, and capabilities, such as a rise in video calls.

Skype's rise is an example of what's called "disruption," a concept formalized by Clayton Christensen in his 1997 book *The Innovator's Dilemma*. Figure 7.1 shows the dynamics of basic disruption. The horizontal axis shows the development of a market over time; the vertical axis reflects how well offerings perform to meet customer needs. The top line shows incremental growth of existing capabilities of an incumbent in a market, called a *sustaining strategy*. The bottom line represents the offering of a newcomer, who provides a lower performing and cheaper solution to address overserved customers.

Although "disruption" is a word with general connotations, Christensen's use of the term has a very specific meaning. Ultimately, disruption is about a competitive response: incumbents striving to sustain an existing business disregard the cheaper, lower quality offerings of newcomers. Then, over time, those newcomers evolve and eventually directly impact existing industries.

JTBD, it turns out, stands at the center of Christensen's theory of disruption. When any business first starts, they focus on solving people's problems to generate demand. But then, as businesses mature, they shift their focus to sustaining the business and stop evolving. A return to jobs thinking, according to Christensen, is the general antidote to disruption.

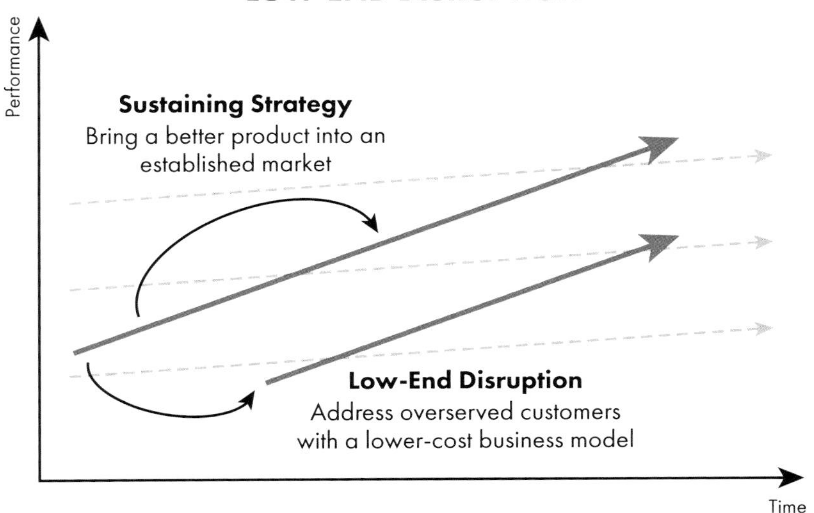

FIGURE 7.1 The dynamics of disruption were outlined by Clayton Christensen: newcomers upend incumbents with cheaper, lower performing offerings.

Disruption is a cautionary tale of how good management—maximizing short-term profits at the expense of developing new solutions customers value—may cause a company's demise. JTBD shifts the focus from products and services to making customers successful in getting the job done.

With JTBD as a North Star, companies are reminded to constantly develop and redevelop their offerings in a way that customers truly value. This chapter looks at ways that organizations can continue to create value through the JTBD lens.

PLAY ➤ Survive Disruption with JTBD

To guard against disruption, Maxwell Wessel and Clayton Christensen propose a simple, straightforward way to look at disruptive threats. In their article "Surviving Disruption" they write:[1]

1. Maxwell Wessel and Clayton Christensen, "Surviving Disruption," *Harvard Business Review* (December 2012).

> Identifying what jobs people need done and how they could be done more easily, conveniently, or affordably is what enables a disrupter to imagine how to improve its product to appeal to more and more of your customers. If you can determine how effective or ineffective the disrupter is likely to be at doing the jobs you currently do, you can identify the most vulnerable segments of your core business—and your most sustainable advantages.

The trick is to identify and define the job that newcomers might easily replicate and develop a disruption of your own before it's too late. They suggest a simple exercise to examine the dynamics of potential disruption using a simple diagnostic approach with JTBD at the core of the analysis.

STEP 1 ➤ Determine the strengths of the disruptor.

First, identify the relevant jobs a potential disrupter gets done. Start with your main job and consider how else people might get it done. Then list the main advantages and disadvantages of the disrupter.

STEP 2 ➤ Identify your own company's relative advantages.

Next, list the jobs your offering currently gets done that overlap with the low-end competition. Describe each briefly from the customer's perspective. Include details about the circumstances and conditions that are most relevant for getting the job done.

STEP 3 ➤ Evaluate barriers.

Examine the conditions that would help or hinder the disrupter from co-opting your current advantages in the future. Determine whether each job core of your offering is easier or harder to disrupt.

To guide the technique, capture your analysis on a simple canvas, such as the one shown in Figure 7.2 from Wessel and Christensen's article. This exercise works well in a small group with team discussion. Make it interactive by using sticky notes and flipcharts to encourage input from everyone.

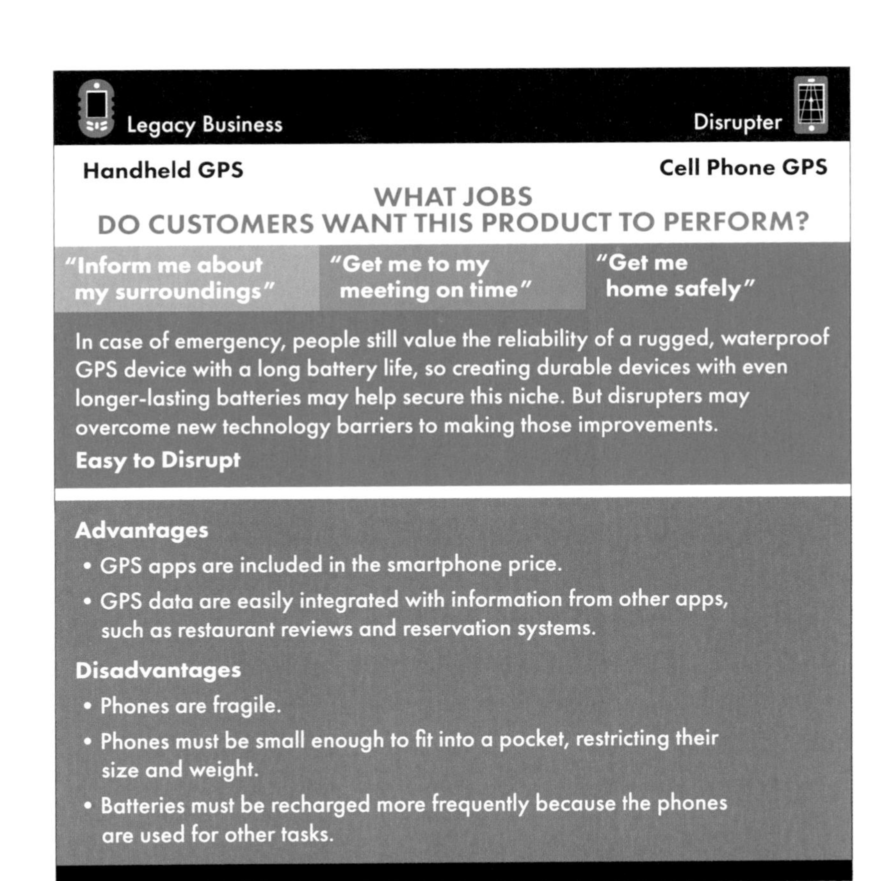

FIGURE 7.2 A quick analysis of JTBD along with advantages and disadvantages provides insight into how to survive disruption.

Notice in this example that the formulation of jobs is more of an imperative a customer may have rather than an objective. The statements use the first person, e.g., "Get me home safely" and "Get me to my meeting on time," as if a fictional person were making a demand on the provider.

To be consistent with your other JTBD efforts, use the framework and language outlined in Chapter 2, "Core Concepts of JTBD." For instance, looking at the above example of a GPS, you could reformulate the elements of getting the job done as follows:

Main job: Navigate to a specific location

Job performer: Individual traveler

Top needs:

- Increase the knowledge of surroundings
- Minimize the chance of being late
- Minimize the risk of unsafe conditions along the way

Primary circumstances:

- When traveling home
- When traveling to a meeting at a specific time
- When the job performer is concerned about safety

Also keep in mind that by changing the main job, job performer, or the solutions compared, you'll get a different picture of disruption. For instance, if you compare a cell phone GPS to the GPS system a ship captain might use, you'll find the lower performing option doesn't reach the same desired outcomes. That's why GPS providers like Garmin have a healthy business in marine navigation and aviation: the jobs to be done in those spaces have not been disrupted with cell phone GPS.

In the end, incumbent companies should not overreact to disruption by dismantling a still-profitable business. Instead they should strengthen their relationships with core customers while also creating a new team focused on the growth opportunities that arise from the disruption.

LEARN MORE ABOUT THIS PLAY:

SURVIVE DISRUPTION WITH JTBD

Maxwell Wessel and Clayton Christensen, "Surviving Disruption," *Harvard Business Review* (December 2012).

> This article details the authors' technique for identifying potential disruption. They directly connect JTBD thinking with disruption in a causal way. Their approach is by no means scientific, but rather provides a light diagnostic thought exercise for insight into potential disruption.

This new capability should focus on a getting a job done quicker, cheaper, with less skill, etc.

PLAY ➤ Strategize Around JTBD

Growth itself is not strategic: all businesses have an imperative to expand. It's *how* an organization chooses to grow that defines a strategy—the unique set of interlocking decisions on what *to do* and what *not to do*. Accordingly, not all strategies are the same, and a company must decide which type to pursue.

Several tools have emerged over the years to help distinguish strategic options. The Ansoff model, for instance, is one of the oldest frameworks in business strategy. The matrix contrasts two dimensions along which to view strategy: the newness of an offering and the newness of the market. The result is four quadrants:

- **Market penetration** is a strategy whereby the organization grows by increasing market share. Aggressive sales, price decreases, and product refinements characterize this approach.
- With the **product development** strategy, a business grows by creating new products and services targeted at existing customers.
- **Market development** allows a company to expand by reaching different customer segments or expanding regions.
- With a **diversification** strategy, growth is achieved by introducing new products to new markets.

Figure 7.3 shows the Ansoff matrix and four primary strategy types from this perspective. Note that this matrix focuses on product strategy, not company strategy. In other words, a company with many offerings may have multiple strategies going on at the same time. Using McDonald's as an example, each quadrant contains a representative product or service that reflects how the fast food chain has grown.

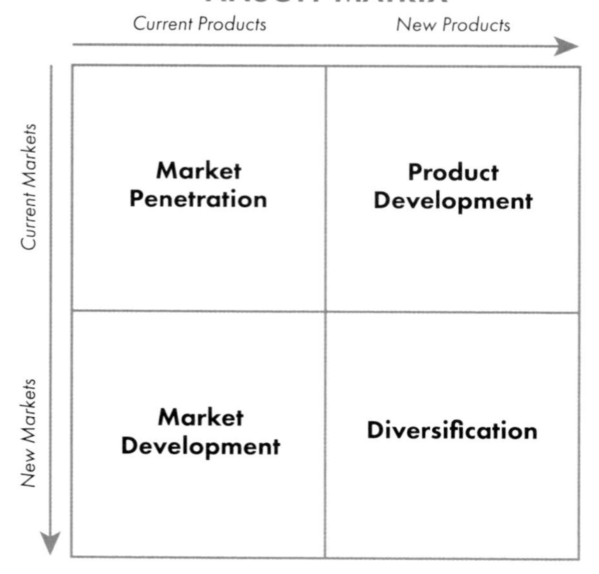

FIGURE 7.3 The Ansoff matrix provides four different strategies that a company can pursue.

Other such matrices look at market share and growth. The infamous BCG growth-share matrix, shown in Figure 7.4, is a tool for assessing a portfolio of products from a strategic perspective. There are four types of offerings as follows:

- **Cash cows** are where a company has a high market share in a slow-growing industry.
- **Dogs** are offerings with low market share in a mature, slow-growing industry. These units typically break even and drain the overall profitability of the company.
- **Question marks** are offerings with a low market share in a high-growth market. They must be analyzed carefully to determine whether they are worth the investment.
- **Stars** are units with a high market share in a fast-growing industry. They are graduated question marks with a market- or niche-leading trajectory.

FIGURE 7.4 The infamous BCG growth-share matrix presents a typology of four different strategies.

Modern approaches to distinguishing strategy extend classic models like the Ansoff matrix or BCG growth-share matrix. In their book *Your Strategy Needs a Strategy*, authors Reeves, Haanaes, and Sinha propose three different axes on which to view strategy based on three questions.[2]

- Do we shape our industry?
- Is it predictable?
- How harsh is our environment?

The resulting strategy palette, as they call it, yields five different strategic types, shown in Figure 7.5.

2. Martin Reeves, Knut Haanaes, and Janmejaya Sinha, *Your Strategy Needs a Strategy* (Boston: Harvard Business Review Press, 2015).

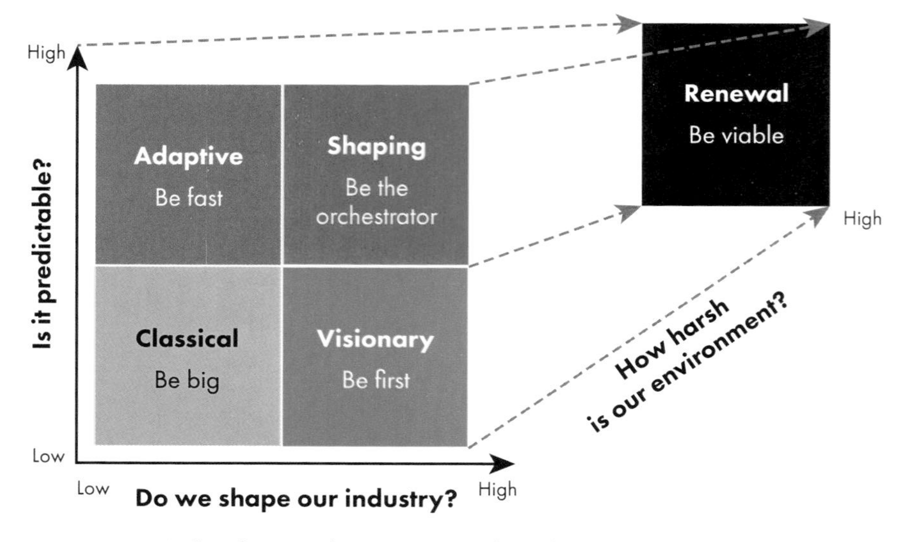

FIGURE 7.5 A classification of strategy types along three dimensions.

Such modules seek to create a typology of strategy, but from a decidedly product-centric view of the market. How can a business position itself? How will it approach a market? What technologies are needed? They look at the marketing from the inside-out.

The JTBD perspective offers a new way of looking at strategy from the outside-in, from the customer's perspective. Focusing on the job allows organizations to maintain a constant strategic imperative—get the customer's job done—even as technology changes.

JTBD Growth Strategy Matrix

Developed by Tony Ulwick and Strategyn, the growth-share strategy matrix presents different strategic approaches based on JTBD (see Figure 7.6).[3] The idea is straightforward: by knowing which solutions get the job done cheaper and quicker, organizations can use JTBD to achieve more predictable growth.

3. Tony Ulwick, "The Jobs-to-Be-Done Growth Strategy Matrix," *JTBD+ODI* (blog), January 5, 2017, https://jobs-to-be-done.com/the-jobs-to-be-done-growth-strategy-matrix-426e3d5ff86e

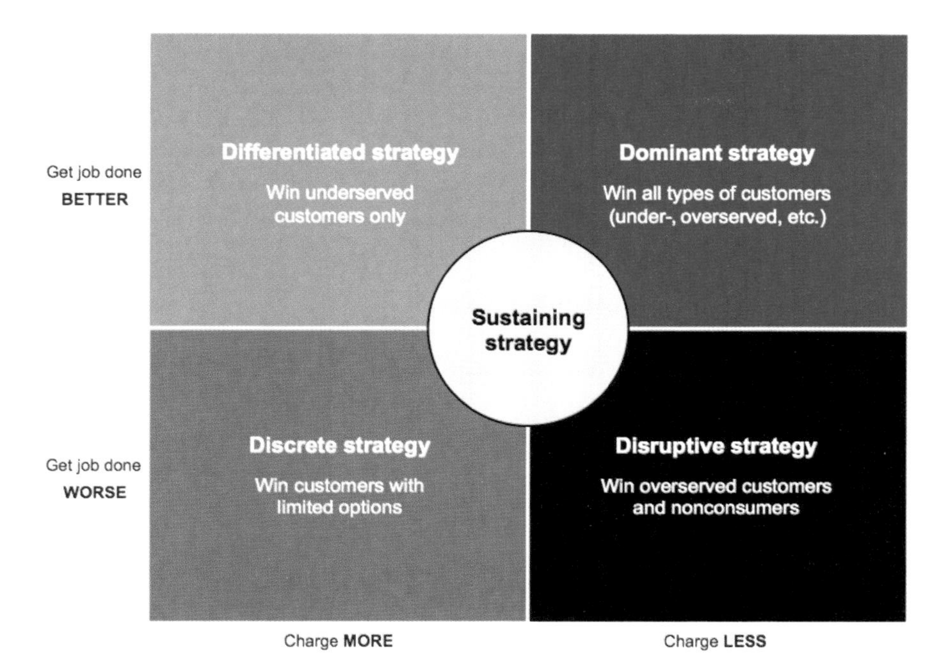

FIGURE 7.6 The growth strategy matrix developed by Tony Ulwick and Strategyn provides a topology of strategic approaches based on JTBD.

The matrix is based on the observations that offerings that win in the marketplace get a job done better and/or cheaper. Accordingly, on the one side, the matrix evaluates how well an offering gets a job done compared to existing solutions on the market—better or worse. On the other, the matrix considers the cost of the offering, whether it will be more or less expensive.

This yields four quadrants, which correspond to types of customers to target relative to their JTBD:

- A **differentiated strategy** targets underserved customers. Offerings are better than existing alternatives and more expensive. Think: the Nest thermostat, Nespresso's coffee machines, and Whole Foods' organic food products.

- A **dominant strategy** appeals to all types of customers with a better-performing, less-expensive offering. Examples include UberX and Netflix.
- A **discrete strategy** is positioned to serve customers with limited options, as well as nonconsumers. Examples include roadside stops on interstate highways, stadium concessions at sporting events, and ATMs in remote locations.
- A **disruptive strategy** aims at overserved customers and nonconsumers. Examples include Google Docs (relative to Microsoft Office), TurboTax (relative to traditional tax services), and eTrade's online trading platform.

A fifth category—a **sustaining strategy**—occupies the middle region on the matrix. This represents offerings that get a job done slightly better or slightly cheaper. Such a product will likely fail to attract new customers. This is a poor strategy for a new market entrant, but it may help an incumbent company retain existing customers.

Here's how the growth strategy matrix works and the steps you can take to craft a job-based strategy:

STEP 1 ➤ Segment customers by JTBD.

Use the techniques described in Chapter 4, "Defining Value," to determine whether there are underserved or overserved customers or both. This entails surveying job executors to find important but unsatisfied desired outcomes to pinpoint unmet needs.

STEP 2 ➤ Decide on a strategy.

Using the matrix, decide which strategy type to pursue. Remember to focus on an individual offering or a suite of products. Your company may then pursue different growth strategies for different offerings. Like the Ansoff model or BCG growth-share matrix, a given company may have offerings in different quadrants.

STEP 3 ➤ Determine solutions that get the job done.

Decide which products and services and the capabilities of each that will appeal to the segment you're targeting. There are four different primary ways to grow a market based on JTBD:

- **Get more steps done.** Review your job map and consider how to get more steps done.
- **Get steps done better.** Compare getting steps done to competing solutions and create ways to accomplish the desired outcomes better.
- **Get related job steps done.** Look at related job and consider how to get them done or integrate them into your solution.
- **Ideate and test solutions that address the job in a strategic way.**

When designing a solution, take into account emotional and social factors, as well as the circumstances of getting the job done. It's unlikely that focusing on getting the functional job steps done will alone result in an attractive offering. You also need to consider how to make a compelling product or service that consumers truly desire. Test your ideas and concepts to refine the product-market on functional, social, and emotional levels.

STEP 4 ➤ Craft a value proposition and create marketing campaigns.

Finally, use JTBD and language to create communications and messages that resonate with the targeted segment. Reflect needs in the messaging. Build a marketing strategy around the job and unmet needs for each segment. Here, too, emotions and social aspects come into play: focusing on the functional steps of the job to be done won't be enough to form a persuasive message.

Keep in mind that the position of a product may shift over time as offerings and market needs change. Additionally, a company may consciously move positions of offerings by going downstream or upstream. For instance, Uber first came onto the market with a differentiated strategy: Uber Black got the job done better, but was more expensive than traditional taxi cabs. But it moved to a dominant strategy with UberX, and then to a disruptive strategy with Uber Pool.

Or a product may start out with a disruptive strategy—getting the job done worse but at a lower price. Then, over time, as the offering improves, it may start to get the job done better, reaching a dominant strategy, such as with Skype. Any analysis done with the matrix is dynamic and reflects a point in time.

If JTBD is a core factor that explains market disruption, basing your offering strategy on JTBD helps give predictability into disruption and overall long-term success. Although present here in a few simple steps, using the growth strategy matrix requires rigor and mastery of both JTBD and strategy.

LEARN MORE ABOUT THIS PLAY:
STRATEGIZE AROUND JTBD

Tony Ulwick, "The Jobs-to-Be-Done Growth Strategy Matrix," *JTBD+ODI* (blog), January 5, 2017, https://jobs-to-be-done.com/the-jobs-to-be-done-growth-strategy-matrix-426e3d5ff86e.

> This article explains the theory behind the matrix and how to put it into practice. There is a link to a webinar recording explaining growth strategy in more detail, framed by a broader conversation of JTBD theory. Also see Ulwick's book, *Jobs to Be Done* (2016), for more information on the growth strategy matrix.

PLAY ➤ Organize Around Jobs

In the last six or so companies I've worked for, I've been through seemingly dozens of reorgs. Perhaps you've experienced this, too: every year or so the company decides to change its structure.

But are reorganizations (reorgs) effective? A 2010 Bain & Company report found that only one-third of organization restructuring provided any value, and, in fact, many destroyed value.[4] For sure, the energy put into reorgs can be high—energy that might be better spent elsewhere. In the end, many reorgs are a futile attempt at addressing deeper strategic issues.

The problem is that reorgs focus too much time on hierarchy and lines of reporting. On paper, who reports to whom shifts around—at times dramatically—while, in practice, the work of employees might stay the same. As a result, there is often little difference in the net effect of a reorg. Just consider the before-and-after view of a company illustrated in Figure 7.7.

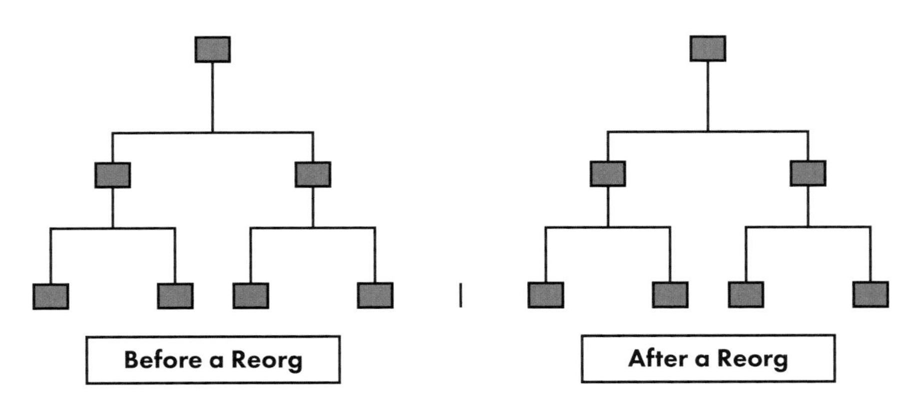

FIGURE 7.7 There is often little difference before and after a reorg.

To avoid the resulting silos inherent in purely hierarchical organizations, some companies have implemented a so-called matrix organization,

4. Marcia Blenko, Michael Mankins, and Paul Rogers, *Decide & Deliver* (Boston: Bain & Company, 2010).

shown in Figure 7.8. This adds another dimension to the structure, and employees have dual reporting relationships. For instance, they may report to both a functional manager and a product line manager.

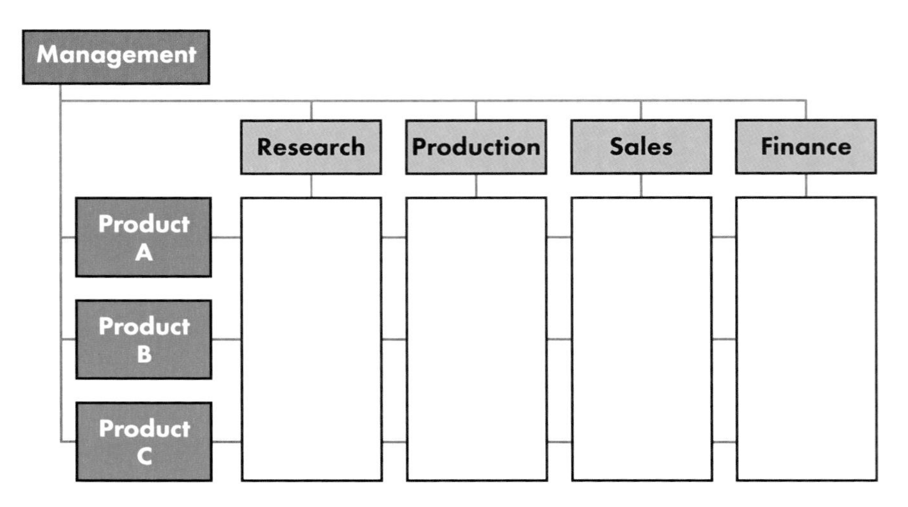

FIGURE 7.8 A matrix company is structured around two organizational dimensions, e.g., functional and product-based.

In another approach, Spotify famously introduced additional dimensions of organization to help their Agile teams be more flexible.[5] Figure 7.9 shows how teams are first divided into *tribes*, or different product lines, led by a product owner, and arranged into *squads*. Then, individual contributors can be organized into *chapters* across squads, as well as into *guilds*, which span tribes.

The Spotify model better represents how people actually communicate inside of companies as a web of connections and communication. Yet even this approach still doesn't ensure that teams will be focused on customer needs. There is nothing inherently customer centric about it.

5. See Henrik Kniberg and Anders Ivarsson, "Scaling Agile @ Spotify with Tribes, Squads, Chapters & Guilds" (white paper, October 2012), https://blog.crisp.se/wp-content/uploads/2012/11/SpotifyScaling.pdf

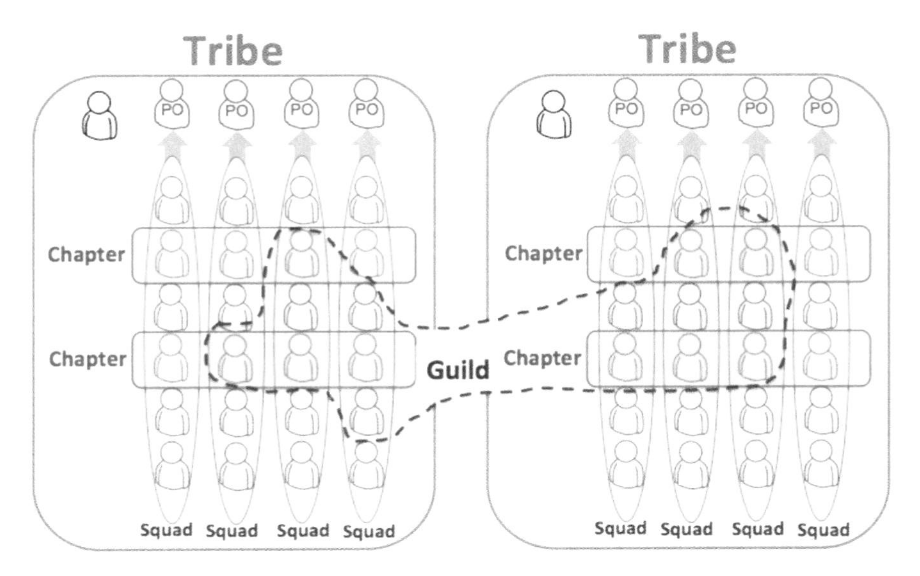

FIGURE 7.9 Spotify's innovative approach to company structure combines multiple layers of organization beyond a typical matrix.

In order to better align teams to customer needs, an alternative is to make JTBD one of your organizing dimensions. This puts a focus directly on customer-centered thinking in a way that is inherent to the company structure. Consider how Clayton Christensen and his coauthors put it in their book *Competing Against Luck*:[6]

> Through a jobs lens, what matters more than who reports to whom is how different parts of the organization interact to systematically deliver the offering that perfectly performs customers' jobs to be done. When managers are focused on the customer's job to be done, they not only have a very clear compass heading for their innovation efforts, but they also have a vital organizing principle for their internal structure.

6. Clayton Christensen, Taddy Hall, Karen Dillon, and David S. Duncan, *Competing Against Luck* (New York: HarperBusiness, 2016).

Instead of working within functional silos, teams can see what matters to customers: getting the job done.

STEP 1 ➤ Cluster jobs into local groupings.

Use your job map to find natural divisions in the job that will correspond to different teams and efforts. For instance, you can look at the steps *before*, the steps *during*, and the steps *after* executing a job as three separate clusters.

Alternatively, if your organization addresses several different related jobs, consider how related jobs affect and influence each other. In the end, you should have a handful of job groupings that could logically form a focus for a given team. The aim is to find big jobs to use as an organizing principle.

STEP 2 ➤ Organize around jobs.

First, determine the level at which to organize around jobs. It's unrealistic to start by making jobs to be done the primary organizing vector. Company hierarchies and roles will likely continue being useful to manage reporting lines, finances, and other aspects of the business. Additionally, people external to your company—including your customers—will expect a certain degree of predictability to your company.

Instead, the simplest way to start is to align cross-functional teams or working groups to jobs to be done at a secondary tertiary level, similar to the guilds or chapters in the Spotify model. This will give you a mix of traditional hierarchies and customer-centric alignment to jobs.

STEP 3 ➤ Set success metrics and measurements.

Finally, give the new teams the mission of "owning" customer success in getting that job done. To the degree possible, empower them not only to understand the job but also to devise ways to solve for it. Also determine the metrics that will be used for success for each job.

While you may not be reorganizing people and reporting lines around jobs, you can orient projects and work in general to customer objectives. This will greatly increase your company's ability to understand the value they create from the outside-in, and to ultimately provide better products and services.

Once you define the jobs your customers are trying to get done and how you want to create a solution to meet their needs, you also need to ask the question, "How does our company need to organize its capabilities in order to provide an appropriate offering and experience?"

Consider the example of Intercom, providers of a messaging platform that helps companies stay in direct contact with their customers. Since launching in 2011, the startup has grown significantly. Initially, the company was organized around JTBD.

They saw a siloed mess of tools to connect with customers: one tool when you wanted to message your customers, a different tool when your customers wanted to message you, and a different one again if you wanted to ask a question. And yet another one if you wanted to send a product update. Sure, it would be possible to integrate various tools via APIs and middleware, but the result would not be seamless (see Figure 7.10).

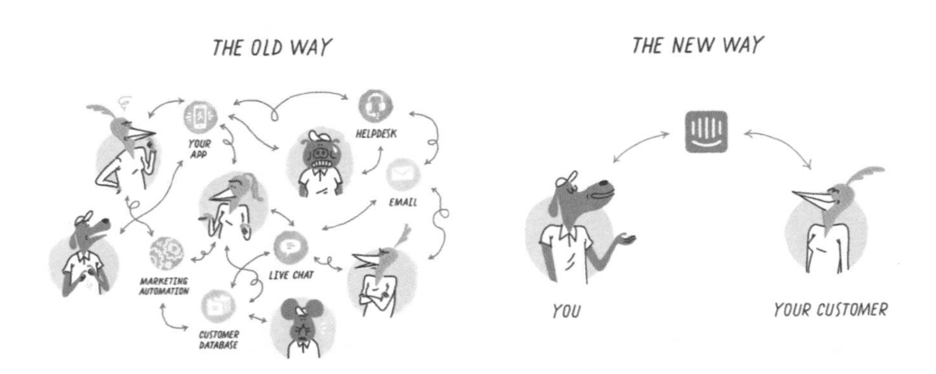

FIGURE 7.10 Intercom used JTBD to sort out a complex set of capabilities with existing tools and simplify with their all-in-one solution while still providing clarity.

However, because Intercom was consciously looking to integrate various capabilities, they quickly realized that they needed to clarify the functions of their software. To do this, they turned to JTBD to help them organize their capabilities. They have used JTBD ever since to bring focus and clarity to their business.

When Intercom first launched, the whole package came bundled together. You got all of it—the support, marketing, and product feedback tools—for a price that varied, based on the number of active users you had in your database. The company broke its offering into four separate capabilities, each targeting a job that their target users wanted to get done.

- **Acquire new customers.** Chat with visitors on your website to help them become customers with Intercom.
- **Engage with existing customers.** Intercom lets its users build relationships with their customers through personalized messages.
- **Learn about customers.** Intercom makes it easy for product managers and research teams to send highly targeted messages to customers, based on what they've done (or haven't done) inside the product.
- **Support customers.** The software solution is also a tool for support teams who need to deliver customer support without traditional "tickets." Instead, teams can distribute support requests and reporting throughout the company.

While Intercom maintained the usual functional roles—marketing, sales, product, etc.—they also organized around the job to be done in terms of daily work and activities.

Or consider how the bank at USAA, a financial services and insurance provider to military veterans in the U.S., is organized around what they call "experiences." Previously, the bank had typical product-based P&L centers reporting up to the CEO—one for checking accounts, one for credit cards, one for auto loans, one for home loans, and so forth. Around 2016, they got rid of these silos and organized around the experiences their members have, closely resembling a job to be done.

As a result, there are no more *product owners* at USAA Bank, but *experience owners* reporting right up to the president. For instance, now there's a *daily spending* experience owner, who is responsible for not only spending from the checking account, but also on the credit card. In jobs terms, we could say that experience owners are responsible for everything needed to help customers *control personal spending on a daily basis*. In the end, USAA Bank institutionalized an organizational model to align to the needs of the people they serve.

In another example, a large publisher I once worked for did something similar. They found four logical divisions in their offerings that correlated loosely to customer jobs to be done and organized offering teams around those. Like the Intercom example, we had typical role types and reporting lines, but efforts and projects were guided by alignment to categories of jobs we addressed in our offerings. For instance, the sales playbook had sections with materials organized around jobs so that salespeople could pitch how we could get a job done, and not try to sell features.

Perhaps the biggest challenge in becoming a customer-centric company is creating a structure that enables customer-centricity. Companies that are committed and obsessed with the customer outperform competitors. JTBD provides a much-needed view from the outside-in to effect organizational change.

LEARN MORE ABOUT THIS PLAY:
ORGANIZE AROUND JOBS

Clayton Christensen et al., "Integrating Around a Job," Chap. 7 in *Competing Against Luck* (New York: HarperBusiness, 2016).

> Christensen highlights the importance of organizing processes and functions around the JTBD in this chapter. He provides many examples of organizations that have done this, such as the Mayo Clinic, Southern New Hampshire University, and more. Structuring an organization around JTBD provides competitive advantages, the author claims.

CREATING SHARED VALUE

After World War II, U.S. corporations assumed a general retain-and-reinvest approach to business finances. They put earnings back into the company, benefiting employees and making the firm more competitive.

This general strategic approach gave way to a downsize-and-distribute posture in the 1970s. Reducing costs and maximizing financial returns, particularly for shareholders, became a priority. The widely held belief was that profit was good for society: the more companies could earn, the better off we all would be.

Unfortunately, this policy has not made America more prosperous.[7] Since the 1970s, American workers have been working more and making less. At the same time, shareholder value in the form of dividends and CEO wages has increased. It's no wonder that trust in corporations is at an all-time low. We're seeing businesses increasingly blamed for many of our social, environmental, and economic problems.

But the balance is shifting, in particular through initiatives around what's called "shared value," a term coined by strategy experts Michael Porter and Mark Kramer in their landmark article, "Creating Shared Value."[8] No longer can companies operate at the expense of the markets they serve. They write:

> A big part of the problem lies with companies themselves, which remain trapped in an outdated approach to value creation that has emerged over the past few decades. They continue to view value creation narrowly, optimizing short-term financial performance in a bubble while missing the most important customer needs and ignoring the broader influences that determine their longer-term success.

7. See William Lazonick, "Profits Without Prosperity," *Harvard Business Review* (September 2014).
8. See Michael Porter and Mark Kramer, "Creating Shared Value," *Harvard Business Review* (January–February 2011).

Shared value explicitly links revenue to creating social benefit. This, in turn, provides a competitive advantage back to the organization. It's a win-win approach.

Shared value goes beyond social responsibility. It's not about donating to charity. Instead, every time a customer interacts with a company, shared value creates value for both society and for companies. It touches the heart of corporate strategy.

The notion of shared value necessarily means that organizations need to conceive their value proposition in a way that takes people's needs into account. Chief among these is a deep understanding of human needs. For instance, in a video interview Porter advises:

> Figure out what your product is and what your value chain is. Understand where those things touch important social needs and problems. If you're in financial services, let's think about "saving" or "buying a home"—but in a way that actually works for the consumer.

If the share value movement compels businesses to look at strategy from the customer's perspective, JTBD provides a framework for doing just that. For instance, if a financial provider needs to understand "saving for a home" in a way that is compatible with consumer expectations, that provider can model objectives and needs using the JTBD framework.

With shared value, companies need to determine how they fit into the customer job, not the other way around. For instance, a home-buying service might promote healthier lifestyles by coordinating home listings with information about neighborhood walking trails. But potential saved costs of walking could also be included in determining what you can afford. Perhaps the system could show how much money would be saved by reducing gas expenses or getting rid of a car altogether.

With shared value in mind, the strategic aspiration of the company becomes even broader: it's more than just about buying a home or even settling into a home—it's about creating a healthier, environmentally friendly lifestyle when buying a new home.

PLAY ➤ Expand Market Opportunities

In my previous book, *Mapping Experiences* (O'Reilly, 2016), I briefly told the story of a conversation I had with a director of sales at a previous company. He was participating in a strategy workshop I was facilitating. I assumed we'd be considering growth opportunities around customer needs and market feedback. The sales director had a slightly different perspective.

"We have to figure out how to get customers for all they are worth," he explained while wringing an imaginary towel. "If the towels get dry, you just squeeze harder." Given his short-term quotas, his position was understandable, but inappropriate, I thought. Sure, share-of-wallet matters, but that wasn't our focus for this effort.

Markets aren't just a collection of consumers to exploit. Customers are an organization's most valuable asset. The problem is that traditional segmentation and the general view of customers is based on demographic attributes. Consider what Clayton Christensen and coauthors Scott Cook and Taddy Hall had to say in their article "Marketing Malpractice:"[9]

> The prevailing methods of segmentation that budding managers learn in business schools and then practice in the marketing departments of good companies are actually a key reason that new product innovation has become a gamble in which the odds of winning are horrifyingly low.

JTBD again offers an alternative. Jobs define a market—a group of people with a similar objective they want to accomplish. Business leader Rita Gunther McGrath believes that in the future, markets should be seen in terms of what she calls "arenas based on JTBD." She writes in her bestselling book *The End of Competitive Advantage* (2013):[10]

9. Clayton Christensen, Scott Cook, and Taddy Hall, "Marketing Malpractice: The Cause and the Cure," *Harvard Business Review* (December 2005).
10. Rita Gunther McGrath, *The End of Competitive Advantage: How to Keep Your Strategy Moving as Fast as Your Business* (Boston: Harvard Business Review Press, 2013).

> The driver of categorization will in all likelihood be the outcomes that particular customers seek ("jobs to be done") and the alternative ways those outcomes might be met. This is vital because the most substantial threats to a given advantage are likely to arise from a peripheral or nonobvious location.

What's more, since technology can change and jobs are stable, it's strategically advantageous to define your market around the job. For instance, vinyl records and cassettes gave way to CDs, which, in turn, were superseded by MP3s and eventually streaming music services. But the job stays the same: *listen to music*. Just because a technology becomes obsolete doesn't mean the market has to change. In other words, instead of defining markets as "people who buy CDs," define it as "people who listen to music."

More than that, a jobs-based approach can help you expand your view of the market. The way to do so is to take a broader look at getting more jobs done or considering the next level up in the jobs hierarchy. In other words, focus on the progress that people are trying to make in their lives as they seek solutions to fulfill their needs.

Here's how to expand your strategic view of the market:

STEP 1 ➤ **Look at the progress that people want to make.** Consider your main job and explore aspirations from that position. Through interviewing job performers (see techniques in Chapter 3, "Discovering Value"), learn about the progress that people want to make, which typically involves getting multiple related jobs done. Take the interlocking set of related jobs into account. Then ask "Why?" to expand your field of vision and go further up in the goal hierarchy. Hold a team discussion to explore and understand your customers' aspirations.

STEP 2 ➤ Ask "What business are we really in?"

Charles Revson, founder of Revlon, perfectly encapsulated JTBD thinking when he said, "In the factory, we make cosmetics. In the drugstore, we sell hope." In a similar fashion, pose the simple question "What business are we really in?" to your team and facilitate a conversation. Consider all of the possible different answers from Step 1. Strive to formulate a single answer to the question that reflects the progress people really want to make in their lives, but you may have multiple answers.

STEP 3 ➤ Reframe your offering.

Now, consider your current offerings. How do they address the higher-level aspirations of customers? What's missing? What needs to be true in order for you to expand your business? Design and redesign all aspects of your offering to address the higher aspiration—from products and service design, to marketing messages, to overall portfolio strategy.

Consider the recent story about the growth of Airbnb that shows how the young company has already expanded their business imperative and extended their market outlook. Instead of looking at the accommodation booking solution as the product, they looked at the trip or journey a customer was taking as the product.

In an interview in *Forbes*, Airbnb designer Rebecca Sinclair discussed how they used design thinking and journey mapping to change their point of view:[11]

> We started to say "the product is the trip" and began shifting our perspective. We could see completely new possibilities in how we thought about which problems to solve and what to build... When we realized the product was the trip, we started to see Airbnb as a lifestyle company that could believably extend into more aspects of the trip, like Airbnb Experiences.

11. See Emily Fields Joffrion's interview with Rebecca Sinclair in: "The Designer Who Changed Airbnb's Entire Strategy," *Forbes* (blog), July 9, 2018.

As a result, Airbnb has introduced new facets of its offering, including Airbnb Experiences. Now, travelers can book tours of a city by locals, cooking classes, museum visits, and more. By moving up from *book accommodations* to *take a trip*, they avoided strategy myopia and expanded their business greatly. Accordingly, the Airbnb offering now addresses several related jobs to be done while taking a trip. Although this example doesn't refer to JTBD specifically, the thinking is the same and is repeatable following the steps outlined previously. Using JTBD, you can look at your company in a whole new light.

LEARN MORE ABOUT THIS PLAY:
EXPAND MARKET OPPORTUNITIES

Clayton Christensen et al., "Marketing Malpractice," *Harvard Business Review* (December 2005).

> This landmark article by several prominent business thought leaders questions traditional ways of market segmentation. The authors point to JTBD as an antidote. Although they provide many examples and case stories, there is little practical information in this article. Understanding jobs-based segmentation starts here.

Also see Alan Klement's *When Coffee and Kale Compete* (2016) for more on JTBD as a measure of progress and focusing on aspirations.

Recap

Those who are familiar with objective-oriented research and design methods, such as Task Analysis, Goal-Directed Design, Contextual Inquiry, and others may struggle to see the difference between these approaches and JTBD. A big difference is the application to business strategy—from defining markets to developing a strategy to creating a future vision of customer value.

More specifically, jobs to be done are a key antidote for disruption. Maxwell Wessel and Clayton Christensen showed how jobs thinking can be used as the basis for a simple analysis of market threats. Comparing an incumbent's offering to competing solutions to get the same job done provides valuable strategic insight.

JTBD is also a strategic driver. The growth strategy matrix, developed by Tony Ulwick, provides a typology of strategies that a company can follow based on JTBD. Similar to other such matrices, there are several options to follow, depending on your market analysis and strategic imperatives.

Companies that aren't aligned toward the customer have more difficulty acting in a customer-centric way. To overcome inherent silos in hierarchical org charts, organize and integrate around the job to be done. While not easy to do at the highest level of reporting lines, it's possible to map jobs to teams and functions as secondary and tertiary levels of structure.

Products that get the job done better than others tend to win in the market. To expand your strategic field of vision, consider how to *move up in the jobs hierarchy*, including targeting aspirations. Have your team ask itself, "What business are we really in?" to find ways to grow.

CHAPTER

8

JTBD in Action

IN THIS CHAPTER, YOU WILL LEARN:

- Full-fledged methods for JTBD
- Recipes for combining JTBD techniques
- How to evangelize and advocate for JTBD

In a company meeting upon his return to Apple in 1997, Steve Jobs declared: "You've got to start with the experience and work back towards the technology." With that, he gave insight into how he was going to turn around the then failing company. His strategy required nothing less than reversing the principles by which software was created and sold.

At the time, Jobs's approach seemed revolutionary. But such thinking is hardly new. As early as 1960, Theodore Levitt discussed the business importance of focusing on customers. In his influential article, "Marketing Myopia," Levitt wrote:[1]

> An industry begins with the customer and his needs, not with a patent, a raw material, or a selling skill. Given the customer's needs, the industry develops backwards, first concerning itself with the physical delivery of customer satisfaction. Then it moves back further to creating the things by which these satisfactions are in part achieved.

Like Jobs, Levitt doesn't just play lip service to serving the market. He advises companies to literally build businesses around meeting people's needs. Being customer centric is not a checklist of activities to work through—it's core to what a company is and does.

The customer-centric imperative is more critical than ever before. Customers have seemingly unlimited choices of providers, giving them real power to switch as needed. Customer-centricity, it turns out, is synonymous with good for business. Just consider some of these findings:

- Research by Deloitte found that customer-centric companies were 60% more profitable compared to companies that were not focused on the customer.[2]
- A Gartner study found that 89% of companies expect to compete mostly on the basis of customer experience.
- Forrester found that customer-driven financial service providers exceeded revenue growth expectations by 30% and nearly doubled stock price expectations.[3]

1. Theodore Levitt, "Marketing Myopia," *Harvard Business Review* (July 1960).
2. "Customer-Centricity: Embedding It into Your Organisation's DNA" (white paper, Deloitte, 2014).
3. Karin Fenty, *The Business Impact of Investing in Experience* (Forrester, April 2018).

Yet, though seemingly straightforward, organizations struggle to truly adopt this mindset. They are stuck in old ways of management and metrics of the past, failing to view the market from the outside-in.

Part of the problem is with the term "customer" itself, which to many people is limited to "consumption." JTBD instead focuses on individuals and the goals that people have independent of a solution, company, or brand. The experience Steve Jobs envisioned wasn't just a better product experience, but a meaningful interaction with a new technology. Likewise, Levitt wasn't merely talking about product satisfaction, but about fulfilling basic needs. Customer-centricity goes deeper than mere consumption and must include an understanding of human motivations.

JTBD is a way to broaden the meaning of customer-centricity in your organization. Though no silver bullet, JTBD provides a common language and understanding of your markets to inform everything from strategy to design, development to marketing, and sales to customer support. It's a consistent engine of inquiry that can drive decision-making at points all along the value-creation cycle.

JTBD Methods

This book presents a collection of individual exercises and techniques that can be practiced separately or in conjunction with each other. They all have a shared focus on the underlying intent that people have when reaching an objective independent of a given solution. Full-fledged methods for applying JTBD to innovation efforts also exist, reviewed below. Each transforms jobs theory into an effective and repeatable innovation practice in its own way.

Regardless of whether you follow an established method or create an ad hoc sequence of JTBD activities to follow, strive to include your teammates in the conversation. JTBD provides a common basis for understanding the needs of your market across the silos of your organization. Getting everyone on the same page increases customer focus, reduces coordination costs, and speeds up decision-making in general.

Outcome-Driven Innovation (ODI)

Tony Ulwick's ODI is arguably the most comprehensive and refined approach using JTBD today. His method represents an end-to-end process for applying JTBD across a company's strategy. The ODI-inspired techniques highlighted throughout this book are my interpretations of the approach based on my own practical experience. I encourage you to learn more about ODI directly from the resources referenced here.

Summarized at a high level, ODI has four phases:

1. **Identify the job.** The main job is a broad functional objective with a portfolio of desired outcomes (i.e., needs) with emotional and social dimensions. Jobs are discovered through in-depth primary research with job performers.
2. **Create a job map.** The job is represented as a process, not a static point in time, captured in a visual diagram to show how it unfolds. The job map becomes a key model to organize insights throughout the process.
3. **Define the desired outcomes.** Needs are seen in relation to the main job, and each main job may have 50 to 150 desired outcome statements that are uncovered through the research.
4. **Quantify the market.** Using a survey to find unmet needs outlined in Chapter 4, "Defining Value," it's possible to pinpoint the market opportunity from a JTBD perspective.

Once completed, the insights can feed into everything from creating a corporate strategy to building a product roadmap and even forming marketing campaigns. Remember that ODI is not a product design method, but rather a market-level view of any business through the

JTBD lens. The impact of ODI targets the highest levels of an organization and can be far reaching.

But with 84 detailed steps, ODI requires extreme rigor to execute properly. And cutting corners doesn't work—the results will be skewed and give false readings. In particular, the last step—quantifying the market—is problematic to get right. The survey used to quantify market need is challenging for participants to take, even with a hefty incentive. Then the analysis requires precise measurements. ODI is difficult to replicate without training, guidance, and practice.

I should know: I've attempted the full ODI process on several projects. My team and I struggled with the amount of time and resources needed to get it right. Quantifying the desired outcomes became problematic due to a variation in wording, the required sample sizes, and the statistical analysis.

I've also worked downstream from an ODI project conducted by Strategyn, Ulwick's consulting company. Unfortunately, their conclusions provided little guidance for my design team, and the project quickly went in many alternative directions. ODI simply didn't live up to its promise in this case.

However, you don't have to complete the entire ODI process to get value. Conducting interviews and creating a job map may be enough in most cases. I find that the job map can stand on its own and provide valuable insight. Critics of ODI overlook this point and tend to focus on the problematic quantification part of the process.

ODI is well documented and accessible. There are a wealth of resources to learn more about, including Ulwick's full-length book, which is available as a download free of charge.[4] Although the name is trademarked and the process is patented, the ODI method can be applied on your projects without prior permission.

4. Get *Jobs to Be Done* free online here: https://jobs-to-be-done-book.com/

Jobs Atlas

In their book, *Jobs to Be Done*, Stephen Wunker, Jessica Wattman, and David Farber recommend creating what they call a *Jobs Atlas*, or an overall look at the landscape of getting a job done. Creating a Jobs Atlas represents an end-to-end method, broken into three broad stages:

Stage 1: Know where you're starting from.

The aim at this stage is to discover the jobs in your field that people are looking to get done. Start by understanding the problem space from the individual's perspective. A key step is to find the *drivers* for getting the job done, or the underlying contextual elements that make a job more or less important for specific customers.

There are three types of drivers to uncover in interviews according to the authors:

- **Circumstances** are the near-term situational factors that influence decision-making, similar to the circumstances outlined in Chapter 2, "Core Concepts of JTBD," of this book.
- **Attitudes** represent key personality traits that might influence how a job gets done, including social pressures and expectations from others.
- **Backgrounds** are the long-term context that affects decision-making, such as environmental factors (weather), work schedules, or unexpected events.

Then you can determine the current approaches that people have adopted to get the job done. To do this, create a scenario storyboard (shown in Figure 8.1) that not only includes the job steps, but also other stakeholders involved, pain points, and the current means of getting each step done.

The pain points, in particular, provide opportunities for you to help the customer get the job done better. In this sense, they resemble unmet needs in the job process.

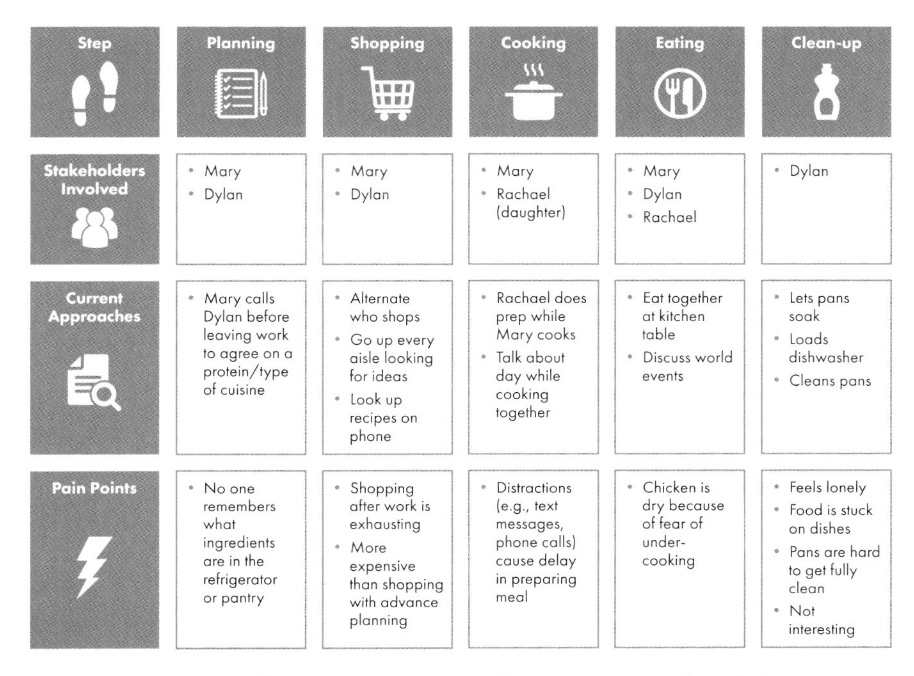

Step	Planning	Shopping	Cooking	Eating	Clean-up
Stakeholders Involved	• Mary • Dylan	• Mary • Dylan	• Mary • Rachael (daughter)	• Mary • Dylan • Rachael	• Dylan
Current Approaches	• Mary calls Dylan before leaving work to agree on a protein/type of cuisine	• Alternate who shops • Go up every aisle looking for ideas • Look up recipes on phone	• Rachael does prep while Mary cooks • Talk about day while cooking together	• Eat together at kitchen table • Discuss world events	• Lets pans soak • Loads dishwasher • Cleans pans
Pain Points	• No one remembers what ingredients are in the refrigerator or pantry	• Shopping after work is exhausting • More expensive than shopping with advance planning	• Distractions (e.g., text messages, phone calls) cause delay in preparing meal	• Chicken is dry because of fear of under-cooking	• Feels lonely • Food is stuck on dishes • Pans are hard to get fully clean • Not interesting

FIGURE 8.1 To get a clear starting point, map the steps in getting the job done (here: *prepare a meal*) along with the stakeholders involved, current approaches, and pain points.

Stage 2: Chart the destination and roadblocks.

After completing Stage 1, teams may want to start developing a solution right away. As much as possible, resist the urge and continue to understand the problem you're solving. The main intent of this phase is to gather the success metrics that customers will use when evaluating your solution. Independent of features or capabilities, focus on the outcomes that people want.

For instance, the maker of a portable digital music player might be inclined to focus on storage and battery life. But using jobs language would instead frame the most important needs that consumers may have as *increase portability of music collection* and *improve ability to listen to music on the go*. The trick here is to identify what people want more of, what they want less of, and where they seek balance.

Finally, Wunker and colleagues recommend looking at obstacles as well. They write: "If a new offering doesn't fit with ingrained behaviors and expectations, customers will be reluctant to change and will look for reasons not to shift to a new solution." Consider these different types of potential obstacles to adoption, even before you've conceived of a solution:

- **Lack of job knowledge:** Customers don't buy what they don't know and understand.
- **Behavior change:** Getting people to change is often more difficult than expected.
- **Multiple decision-makers:** A purchase requires the coordination of several people, often blocking or slowing potential sales.
- **High costs:** Direct and indirect costs may be perceived as disproportionately high.
- **High risk:** Solutions may bring the risk of failure by using them.
- **Unfamiliar category:** Innovative products that define a new category are harder for consumers to grasp and find immediate value in.
- **Missing infrastructure:** Adoption will be slowed if a solution requires a new infrastructure, e.g., electric cars require a network of charging stations.
- **New pain points:** Switching to a new solution may bring more issues than it solves.
- **Hype:** The luster of cool, new products may wear off before adoption is complete.
- **Off target:** Products are incorrectly targeted or not targeted at the right consumers.

The resulting list of obstacles for your situation serves as a guideline to help form a solution. Think of them as requirements that can also be used to evaluate potential ideas as they are developed.

Stage 3: Make the trip worthwhile.

Unlike other JTBD approaches, Wunker, Wattman, and Farber use JTBD to determine the potential value of an innovation to the business in advance. With this, they're able to translate insights into potential revenue.

The aim is to frame markets in terms of jobs, not products. Look at the different segments you've determined from the circumstances, attitudes, and behaviors of job performers. Then ask, will satisfying a job for each have an impact on the company's bottom line? The point is to consider the business value of fulfilling jobs even before you specify the solution.

Overall, their roadmap for customer-centered innovation serves as a practical approach for creating products and services that people really want. Keep in mind that the Jobs Atlas sits within a broader roadmap for innovation, as well. Creating a Jobs Atlas should feel quite practical. It consists of techniques that are straightforward and intuitive to use for a range of stakeholders.

Switch and Four Forces

The Switch technique also provides a guided sequence of steps for applying JTBD. After the in-depth Switch interviews, the Four Forces analysis provides a framework for understanding customer motivations. This insight can then be used to improve existing products or creating new innovation.

Switch doesn't represent a complete innovation method, but it fits well with other techniques, like Design Thinking, Lean, and other existing approaches to innovation. A sequence of steps using Switch might look as follows:

1. **Conduct interviews.** As an overall approach, Switch is rooted in its straightforward interviewing technique. All investigations within this approach to JTBD begin with interviews to understand the underlying motivation to switch ways of working, which, in turn, points to a job to be done.

2. **Find patterns.** Look across interviews and the Forces analysis to find common themes. Pull out the main patterns you find in a summary for your team to review. The aim is to identify the key job motivations that will help you predict why people might hire a solution.
3. **Identify opportunities.** Take the drivers you've identified into a workshop with a mix of participants. First, present the main drivers you've found and have each participant generate challenge questions to address during ideation. Instruct them to begin each with the phrase "How might we...," so they are all in a common format. Then cluster and prioritize the questions across the team to find the most important challenges to tackle.
4. **Ideate solutions.** Guide the group through structured brainstorming activities. Assign the most salient "How might we..." questions separately to individual breakout groups. Each group then devises solutions to address their challenge. Present the final solutions back to the group for discussion and iteration.
5. **Experiment.** Ideas and concepts that emerge from ideation are not ready to be implemented. Experimentation is needed to refine the solution. Find ways to quickly prototype, test, and iterate the proposed solutions.

In the end, complete methods are great to have, but rarely are they followed exactly. There are always local situations that call for variation: time, budget, personalities involved, and desired results all inject variation. By breaking down JTBD approaches into individual plays, as I've done in this book, my hope is to give you the inspiration and confidence to combine techniques into different recipes to serve your specific purpose.

JTBD Recipes

The goal of this book is to give you simple and practical starting points for working with JTBD. You don't have to complete a full-fledged method, like the ones outlined previously, to get value from jobs thinking. By design, the techniques outlined in this book can be practiced individually or together, giving you flexibility to customize your approach.

However, creating a sequence of related activities adds to the strength and power of your research. It's up to you to determine which sequence to follow to fit your situation. It's up to you to combine techniques together in a logical order.

Following are recipes you can use to craft your own program of JTBD research. They are organized around common goals that a team or organization may have, in general. Each begins with scoping the JTBD playing field.

Overall, your aim is to understand the market from the individual's perspective and create a model of the problem space in which you're operating. From there, you can determine the best opportunities to pursue and create an appropriate solution to address people's needs. JTBD offers practical approaches throughout the value creation cycle.

Keep in mind that JTBD is compatible with other methods and processes. For instance, a job map and need statements uncovered during JTBD research can serve as rich input into typical design-thinking activities. JTBD provides a highly structured exploration of the desired outcomes that people have independent of an offering, increasing empathy needed to kick off ideation and solution finding.

That same JTBD research can also feed in the Lean experiments. Use an understanding of the most important jobs that people have in order to prioritize features to test. Then, during Agile sprints, job stories extracted from JTBD inquiry provide a continuity of focus on customer needs during development.

Recipe 1: Launch a New Product or Offering

If you're an entrepreneur or intrapreneur, your aim is to understand first the potential market needs to increase the chances of success. Even if you may be willing to experiment and "learn" your way into a viable solution, JTBD can help identify the most potentially impactful place to start.

Since jobs are independent of technology, a JTBD investigation can be completed even before a product exists. Startups, for example, can systematically investigate the intent and motivation of potential customers even before their solution hits the market.

The aim with this recipe is to ensure that product-market fit from the outset, or the degree to which a product satisfies a market need. JTBD helps focus your efforts on those aspects that result in the highest demand. But by targeting the main functional job first, and then layering in emotional and social factors, your chances of finding an ideal product-market fit increase, and you'll waste even less time testing inappropriate solutions.

A sequence of JTBD techniques for launching a new offering might look as follows:

1. **Conduct jobs interviews.** Even before you have a customer base, you can focus on getting feedback from job performers. Find people who get the main job done and engage them in interviews.
2. **Create a job map.** Even without a product or service on the market, the job can be mapped. A completed map serves as a focal point for discussion to map your assumptions.
3. **Find underserved needs.** Strive to locate the best opportunity from a customer perspective. To keep this activity light, follow Dan Olsen's approach to finding underserved needs outlined in Chapter 4 and found in his book *The Lean Product Playbook*.[5]

5. Dan Olsen, *The Lean Product Playbook* (New York: Wiley, 2015).

4. **Create a value proposition.** Based on the job map and underserved needs, form a hypothesis value proposition to test and refine.
5. **Test assumptions.** Conduct a design sprint or similar activity to form testable concepts and identify the riskiest assumptions. Then devise experiments to test and validate your hypotheses.

From here, you should have significant insight to forge a clear path for moving forward. I recommend following Jeff Gothelf's Lean UX approach, which will guide you through cycles of building, measuring, and learning.[6]

Recipe 2: Optimize an Existing Product or Service

I suspect most readers of this book probably work in situations where there is an existing offering and have a desire to improve their products or services. Product managers, designers, and even developers can use the lens of JTBD to find and prioritize areas of improvement for current solutions already on the market.

Overall, the aim in this case is to find the most impactful enhancements. Below is a sequences of JTBD techniques you can use together to optimize an existing solution:

1. **Conduct jobs interviews.** Interview job performers to find out what they are trying to accomplish. Since there is an existing offering, you can recruit from your customer base, but be careful to avoid talking about your solution during the interview. Instead, focus on the job and how they get it done.

6. See Jeff Gothelf, *Lean UX* (Sebastopol, CA: O'Reilly, 2013).

2. **Compare competing solutions.** Distill need statements from your interviews and compare how well competing solutions get the job done or not. Discuss the competitive landscape with your team and determine your sweet spot. Then take that insight into ideation sessions to come up with ways to get the job done better than others.
3. **Create a consumption journey map.** Diagram the current journey that customers have when interacting with your company. Look at the functional, emotional, and social jobs fulfilled along the way.
4. **Write job stories.** Once you narrow down actionable concepts to implement, create job stories to tie the JTBD research to product development and marketing campaigns.
5. **Create a roadmap.** Use the JTBD framework to guide your plan for improvements. Organize the roadmap around the core job themes you uncovered in your research.

As an alternative to jobs interviews, Switch interviews can also be used to find underlying problems to solve. In fact, you may find it easier to get started with Switch interviews when talking to existing customers. Be careful, however, not to focus only on the purchase decision-making process too much. Also be aware that in B2B situations, interviewing buyers may miss out on end-user feedback.

The output of the above activities can feed directly into standard design and development stages. Bring your team together to ideate the best way to enhance the existing product or service. Use the consumption journey map to pinpoint deficiencies in getting the job done and develop ways to overcome friction and pain points. Bring job stories into design sessions to anchor solutions in real customers' needs uncovered during JTBD research.

Recipe 3: Increase Demand for Existing Offerings

If you're a marketing specialist, consider how jobs thinking can inform your efforts. The aim here isn't necessarily to improve a product or service—although that may be an outcome of your investigation. Instead, the focus for this sequence of JTBD activities is to optimize the awareness, promotion, and communication around an offering. Here's what you can try:

1. **Run Switch interviews.** With existing products, Switch interviews make a great starting point for understanding demand. Use the timeline technique to get to the first moment of thought around why customers switched from one solution to another.
2. **Analyze the Four Forces.** First, identify the push and pull motivations from customers represented in Figure 8.2. What problems do they have? What attracts them to new solutions? Then analyze the habits and anxieties that people have using solutions available to them. Look beyond your immediate solution and strive to uncover the progress that people want to make in their lives.

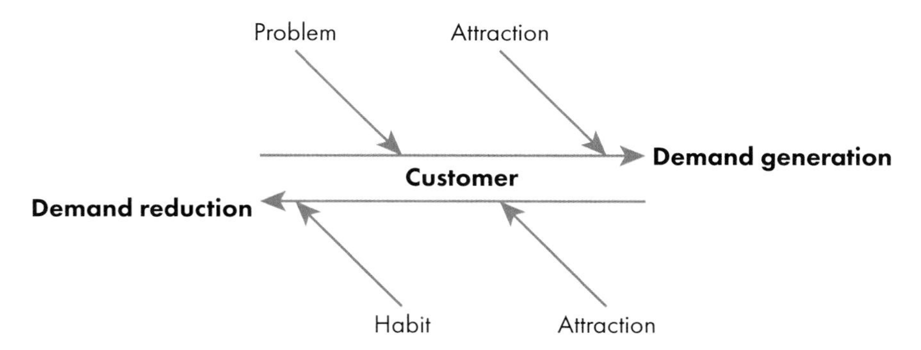

FIGURE 8.2 The Four Forces analysis looks at factors that generate more demand, as well as those that reduce it.

3. **Write job stories.** Capture the insights from your analysis in the form of job stories. Use these to ideate ways to create more demand, as well as overcome factors in demand reduction. Strive to help customers envision how your solution will improve their lives, e.g., with testimonials, user stories, and case studies. Show customers that you understand their struggle for progress. Find ways to demonstrate the advantages of your solution and make it easy for new customers to try it out.
4. **Build a roadmap.** Measures to increase demand may impact both marketing efforts and product development. Create a roadmap to not only set up a sequence of implementation stages but also to align teams around customer goals.

By starting demand generation measures with the job that people want to get done, you'll increase the chances that your message will resonate with them. Focusing on unmet needs further helps differentiate your offering from others.

Recipe 4: Make Customers Successful over Time

Subscription-based services change how companies approach customers. No longer is a one-and-done purchase of a single product relevant. Instead, you have to build an ongoing relationship with the people you serve over time. If you have a business model based on subscription, making customers successful in their job is of paramount importance.

The field of customer success especially focuses on helping people get the most value out of a solution, and JTBD has informed the discipline already. The notion of "desired outcomes," for instance, is a key concept in the field.

If you are a Customer Success Manager, Account Manager, or Sales Executive looking to expand an existing customer base, here is a sequence of steps that you can take to help your people be more successful in getting the job done.

1. **Conduct jobs interviews.** Since you're starting from an existing solution, it will be tempting to discuss your offering. Resist that urge. Focus instead on what customers are trying to get done beyond your product or service. Use the critical incident technique to have interviewees recall a specific situation they were in.
2. **Create a job map.** Distill the steps in getting the main job done in a map. Use this to diagnose your own efforts and where you can make customers more successful. Is it at the beginning of the job process, the middle, or the end? Simple insight into how the job unfolds over time from the customer's perspective helps guide your overall efforts.
3. **Onboard customers successfully.** Don't just onboard people into your product; onboard into the job. JTBD helps you go beyond the boundaries of your solution and find ways to fit into your customers' overall workflow. Craft the sequence of messages and demonstrations of your solution around the main job, and you'll be able to get customers to value quicker.
4. **Maximize customer retention.** After they are onboarded, the aim is to keep customers loyal. You can proactively drive retention by focusing on the customer's job. Devise nurturing campaigns to help them get more of the job done or get it done better. Refer back to the map of your main job to plan communication and education.
5. **Provide relevant support.** Reactively, you'll need to provide support to customers. Your agents can better solve the underlying issues that customer have and address the job to be done directly.

Markets are conversations. It's no longer viable to sell customers products without developing a relationship with them, particularly in subscription-based business models.

Recipe 5: Build a Corporate Innovation Strategy

JTBD theory is based on the simple observation that people don't purchase products, they hire them to get a job done. This powerful heuristic can inform your organization's overarching strategy. Here's how:

1. **Conduct job interviews.** A job interview is a key step in JTBD approaches and can be leveraged in many ways.
2. **Create a job map.** Represent the process of getting the main job done in a simple diagram illustrating the individual steps. Use this to foster discussions about opportunities with your team.
3. **Find underserved needs.** Here, you can also use the circumstantial factors to cluster different outcomes to segment your market.
4. **Create a jobs-based strategy.** Use Tony Ulwick's growth matrix to determine what type of strategy you are targeting. Then consider all of the implications of that strategy type. What needs to be true in order for your organization to deliver against that strategy? You may need new resources or skills or funding types in order for the strategy to be successful.
5. **Organize around jobs.** Peter Drucker once said, "Culture eats strategy for lunch." With this, he exposes the importance of how the organization thinks and behaves collectively. While a simple reorg around jobs won't solve all of your culture and communication problems, it's a step in the right direction. At a minimum, having formal lines of communication that align to your customer's job opens up possibilities that traditional organization schemes get wrong.

Eventually, you'll also want to expand market opportunities. Use jobs thinking to explore what's next for your company's growth. Move laterally to provide service for related jobs. Move up to address high-order jobs and even aspirations.

Bring JTBD to Your Organization

One of the most common questions I get after my talks and workshops on JTBD is, "How do I begin?" Customer-centered advocates want to get started, but face barriers inside of their organization.

The good news is that with increasing frequency, stakeholders are directly requesting JTBD research by name. The bad news is that JTBD requires a change of mindset and behaviors from everyone in the organization. Patience and persistence are required. And even if the will is there, getting started with JTBD in your organization can be a challenge.

Here are some recommendations to consider as you strive to bring JTBD to your organization:

Start small.

Try individual JTBD techniques alone or in pairs on isolated projects. Get a success story quickly that you can use to get time and resources to do more. Don't attempt to follow full-fledged methods, like ODI or the Jobs Atlas, right off the bat. Instead, pilot an effort to learn how JTBD fits into your situation and your organization.

Integrate JTBD into other activities.

Fold JTBD research into other workstreams that are already planned. For instance, if your user research team is conducting an ethnographic study of customers, insert some questions from jobs interviews to collect feedback needed to complete a job map. A job map, in particular, is a powerful way to summarize your overall insights to feed into solution-finding activities, such as design workshops.

Involve others.

In his book on JTBD, Tony Ulwick explicitly recommends limiting decision-making based on JTBD to a small, strategic team. In *Jobs to Be Done*, he states:[7] "What we have learned is that innovation should not

7. Anthony Ulwick, *Jobs to Be Done: Theory to Practice* (Idea Bite Press, 2016).

be everyone's responsibility. It should be the responsibility of a small group of people."

I disagree with this perspective. Modern organizations empower small teams to make local decisions that can have a global impact. They have to have a common perspective on what the organization is doing and where it is heading. This doesn't mean training everyone in the company, but rather instilling a common customer-centric perspective that is engrained in the culture.

JTBD is too powerful and pervasive to be confined to just one team. Instead, strive to involve others throughout your JTBD process and teach them along the way.

Find a champion.

Locate others in the organization who are interested in JTBD, particularly decision-makers. Leverage their interest and enthusiasm to help spread jobs thinking. For instance, get them to sponsor a project that can serve as an example to others in the organization.

Instilling a jobs mindset into your organization can start as a grass-roots effort. But having the weight of a senior leader in the organization will accelerate adoption from above. Bringing JTBD to your organization is both a bottom-up and top-down endeavor.

Know the objections.

If you get pushback from others, be ready with persuasive arguments. Table 8.1 lists some typical objections you may hear from others, the underlying error made in that objection, and suggested counterpoints to make for each.

TABLE 8.1 COMMON OBJECTIONS TO JTBD AND ARGUMENTS AGAINST THEM

OBJECTION	ERROR	ARGUMENT
We don't have time or budget for that kind of research.	JTBD projects take a long time and are expensive.	Working with JTBD doesn't have to be expensive or time consuming. A simple project can be done in a few weeks for about the cost of a marketing survey or usability test. Let's start with some interviews and a job map.
Each department has its own JTBD analysis.	Functional silos work efficiently individually.	Fine. But do they show interaction across channels and touchpoints? Great customer experiences cross our department lines, and we want to create an offering that everyone wants.
We already know all of this.	Implicit knowledge is enough.	Making implicit knowledge explicit is an important part of being customer centric. Also, we won't lose the knowledge when someone leaves. And if someone new joins the team, we can ramp them up quickly. JTBD provides a simple structure for us to capture and organize valuable market insight.
I was in the target group. Just ask me.	Personal past experiences are enough to be customer centric.	Your input will be invaluable in helping us make sense of the job to be done. We also want to ground that perspective with direct feedback from customers: that's where the best insights for growth and innovation are found.
Marketing already does research.	Marketing and JTBD are the same.	Great. But JTBD goes beyond traditional marketing research. We also need to uncover unmet needs and align activities to the main job across departments. With JTBD, we can find our customers' underlying motivations.
We don't need another approach or method—we already focus on customers.	JTBD is the same as other existing methods.	JTBD offers unique value to our organization in several ways. First, JTBD is compatible with other approaches, typically feeding structured insights into underlying needs into existing activities like design thinking, Lean, and Agile. Second, JTBD is broader in scope, looking at market innovation across departments. Third, JTBD isn't owned by any one field. It comes from the business community and can be driven by a variety of role types."

Provide evidence.

Know the benefits of JTBD and be ready to point to success stories. Use the resources and case studies included in this book as a starting point. Find others online and keep a list of relevant resources to show as evidence.

If possible, also find out what your competitors are doing. Search for competitors along with keywords like "jobs to be done" or "JTBD." Showing that others are doing this kind of work goes a long way toward convincing decision-makers.

Create a pitch.

Create a succinct statement that you can readily recite and include the business problems you'll address. Your pitch must be relevant to your situation. Why should a decision-maker invest in a JTBD effort of any kind? Here's an example pitch if you ever get the chance to ride an elevator up with the CEO:

> We'd like to grow beyond our current offerings. By finding our customers' JTBD, we'll have a better understanding of market demand and ultimately the adoption of our solutions.
>
> JTBD is a modern technique to improve customer understanding that more and more companies are using, such as Intel and Microsoft. Some of our competitors are using the technique, too.
>
> With relatively little investment, JTBD provides us with the strategic insight we need in today's fast-changing marketplace. We can get everyone on the same page about what it means to create solutions that customers really want.

In the end, bringing JTBD to your organization doesn't have to be an all-or-nothing proposition. You can get started easily with a small pilot project or by integrating a JTBD technique into an existing project. If people in your organization start to adopt a jobs mindset, you should start hearing them using the language of JTBD.

More importantly, innovation in modern organizations doesn't happen only at the top. Teams following Agile and Lean are smaller and more empowered than in the past, and decision-making is consequently more distributed. When day-to-day decisions are based on jobs, your company will have more focus and alignment so that it can grow.

Defining the Product Roadmap with JTBD at MURAL

By Agustin Soler, Head of Product at MURAL

Tying customers' needs to direct product features is challenging. At MURAL, we try to solve this issue by using jobs to be done. By anchoring customer jobs to be done both globally and locally, we are confident that what we develop is grounded in needs. The process is not perfectly linear, but we tend to follow the steps below consistently.

STEP 1 ➤ Create a job map.

Through research, we created a job map for our main job in collaborative sessions online (see Figure 8.3). The map has eight stages with several steps in each. We also have personas and scenarios that we align to each of the stages, supplementing our understanding of getting the job done.

We can then map the needs to the job map. Having all of this information well organized in one place gives us a valuable perspective. It also forces us to think about the underlying customer needs and problems we are solving.

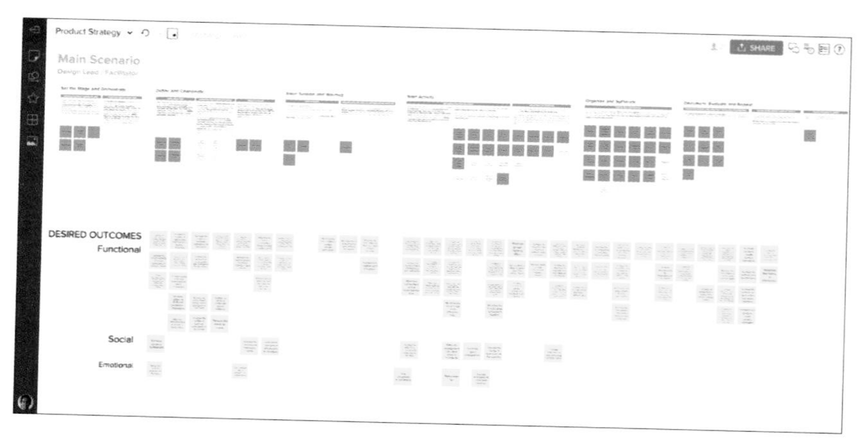

FIGURE 8.3 An excerpt of the job map used at MURAL to plan the quarterly roadmap.

STEP 2 ➤ Decide what to build based on opportunities.
The job map helps us decide what to build each quarter. We find opportunities in the job map by looking for unmet customer needs. These opportunities are prioritized and discussed, and eventually they guide our decision on what to build.

We also follow a simple planning checklist that we've developed over the past few years to ensure that we don't miss anything important. In our experience, it's really key to gather input not only from customers, but also from internal stakeholders inside our company. From this input, we create a development roadmap each quarter focused on what we believe the biggest opportunities are, as seen in Figure 8.4.

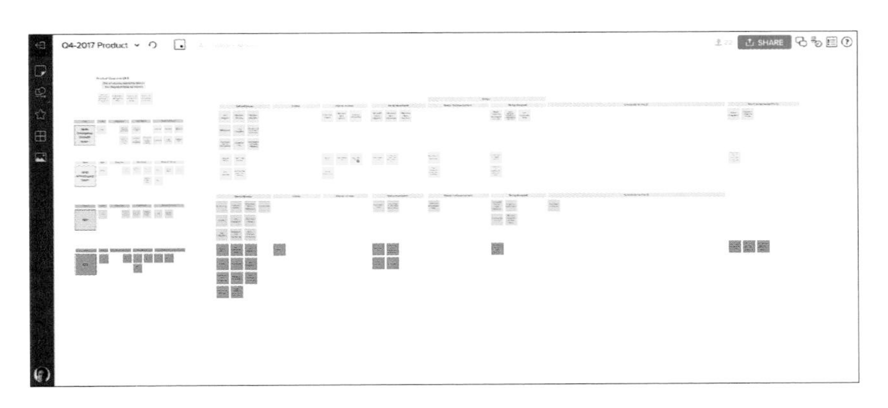

FIGURE 8.4 The quarterly roadmap at MURAL is derived from the job map.

CONTINUES ➤

CONTINUED ➤

STEP 3 ➤ **Research the job.**

We prioritize the opportunities to determine which to work on first each quarter. This typically begins with additional research more specific to the immediate effort. We strive to combine qualitative and quantitative methods to get a broader understanding of the core problem from the individual's standpoint. All of the insights are recorded in MURAL so that we can share them across locations (see Figure 8.5).

FIGURE 8.5 Interview highlights across participants captured in one mural.

For instance, if we want to offer an integration with another application, we'll first interview people to understand about their specific needs. What are the smaller jobs involved along the way? What is their workflow? We also combine insight from interviews with quantitative metrics to inform our point of view. All of this insight feeds into the design process.

STEP 4 ➤ Create job stories.

From our research, we get raw information that we process into job stories, or short statements describing a specific need around specific capabilities. (See Chapter 5, "Designing Value," for more on job stories.) These stories help frame the problems we want to address and figure out how we'll solve them. We discuss, refine, and prioritize these job stories until we get to the ones we think are most important to solve, based on the impact on our users' goals.

Example job stories, in this case for an integration with GitHub.

- When I'm looking at backlog issues, I want to be able to go back to the original source for more context so that I can increase the understanding of the problem I'm trying to solve.
- When I'm conducting a workshop and come up with action items, I want to easily send them to an issue tracker that my team uses so that I can avoid having to change my workflow.

Job stories have the advantage of being portable. They include context from the job map without having to refer to the entire map. This technique enables designers to focus on their specific design challenge with confidence that it ties directly into the overall user experience.

STEP 5 ➤ Design and develop feature.

After we finish the design phase of an issue, it's then ready to be developed. There are many cases in which we first implement a reduced version of the feature just to answer a specific question or see how it works. Developers split the issue into many smaller issues to reduce risk. After they're done working on it, the product manager and designer validate that we built what we expected. Once approved, we do a phased rollout. We first enable the feature in an internal testing version of the app, and then send it to a couple of selected workspaces to gather feedback and iterate.

CONTINUES ➤

CONTINUED ➤

CONCLUSION

JTBD gives us a consistent approach to understand what our customers are trying to accomplish. The common language allows us to tie user needs directly to development efforts. Designers can zoom out to see the big picture in the job map or zoom in to see a specific job story. Developers have comfort that there's a clear rationale behind their product decisions.

In the end, the entire design and development organization is much better aligned. And best of all, we know that our efforts are aligned to what customers really want and value.

Agustin Soler is the Head of Product at MURAL, a company he co-founded in 2012, where he leads a team of designers and developers. In 2016, he received a Master's Degree in Industrial and Product Design from Northwestern University.

Recap

JTBD is a way of seeing people independent of your offering or solution. Coming from the business community, JTBD isn't rooted in any one discipline but rather can be utilized across a company.

Outcome-Driven Innovation (ODI) is perhaps the most complete JTBD method out there today. The full ODI process is difficult to replicate, particularly quantifying unmet needs. However, elements of ODI can be used individually, such as the job map.

Stephen Wunker, Jessica Wattman, and David Farber recommend creating a Jobs Atlas, or comprehensive understanding of the problem you're trying to solve before coming up with a solution. Their approach is part of a larger roadmap for building solutions that people really want.

The Switch technique is frequently paired with the Four Forces analysis. Together, they provide valuable insight into market demand.

This book presents a collection of some of the most popular JTBD techniques in practice today. They can be used individually or put together in recipes. Start with your goal in mind, and use those techniques that will answer your key questions.

Any JTBD effort begins with clarifying your goals and scoping the JTBD landscape. Research is needed to inform your understanding of people's needs. Then you can combine various approaches to fill in your gaps in knowledge. Remember: JTBD is all about understanding the problem and your opportunity before coming up with solutions.

JTBD is compatible with other techniques, such as Design Thinking, Lean, and Agile. You'll not likely replace existing approaches you're using to understand market innovation, but rather supplement them with JTBD. In doing so, be ready with arguments to evangelize and advocate for using JTBD.

Final Thoughts

Purists from the different perspectives of JTBD will likely criticize this book for, well, not being pure. They're right. I've intentionally pulled apart existing methods and remixed them in new ways. In my opinion, JTBD is a perspective, a way of seeing, not a single technique or method.

I also believe that you can't fully comprehend the potential impact of JTBD just from reading a book on theory. I've intentionally gone light on theory in this book in favor of practical applications. Learning by doing is more important than expounding at length about theory alone.

Don't get me wrong: the theory *does* matter. I encourage you to continue reading more about JTBD and filling in the gaps left by this book. But ultimately, you won't adopt JTBD in practice until your team or organization sees its value in action.

I'm also well aware that some critics point out the overlap between JTBD and existing methods. "JTBD isn't new," they say. And they're right, too. As a former product designer, I'm well-versed in methods such as Goal-Directed Design and Task Analysis, among others. I also have direct experience with Voice of the Customer research and other customer experience techniques. I'm fully aware of the similarities between JTBD and these approaches, and the strengths and weaknesses each one has.

Contextual Inquiry as outlined by Hugh Beyer and Karen Holtzblatt, for one, has been particularly influential to how I understand people and the solutions they need. In my opinion, contextual inquiry has the most direct overlap with JTBD. The authors talk about supporting users' "work," and even point out flaws of demographic-based segmentation. But you could replace "work" with "job to be done" in their book *Contextual Design* and end up with nearly the same book.

So what's the difference?

First, JTBD isn't a design method. Goal-Directed Design, for example, specifically targets the design of software interfaces. JTBD is much broader, as I hope I've shown throughout this book. JTBD is a way of viewing your market to accelerate innovation across disciplines.

Second, focusing on people's goals didn't start with Goal-Directed Design, Task Analysis, or something similar. That type of thinking can be traced back to innovators, designers, and entrepreneurs at the beginning of the twentieth century. Historically, JTBD emerged in parallel to other related fields, and there is no evidence that one affected the other.

The important point to remember is that JTBD comes from business, not design or customer experience or product management. This gives JTBD a different sense of importance and weight, as well as makes it potentially more pervasive throughout the organization. JTBD is not "owned" by design or any other discipline for that matter.

In my experience, traditional design and UX practices have largely failed to inform businesses what do in order to become customer centric. Practitioners of human-centered design will be the first to admit that their organizations don't listen to them. They long for that proverbial seat at the table. To be fair, things are changing, as evidenced by efforts like IBM's multiyear rollout of Enterprise Design Thinking. But overall, design only is not enough to shift the organization, practices, and mindset of entire organizations.

And finally, JTBD provides a concrete unit of analysis and common language that spans departments and roles. There's a clear sequence to the focus on JTBD: start with the main functional job and then layer on emotional and aspiration considerations. JTBD makes focusing on people tangible.

So, yes, the human-centricity inherent in JTBD is not new. But it does represent a new opportunity for us to change the direction, practices, and collective mindset of an organization. In the end, JTBD has the potential of becoming a substrate that runs through an organization, as seen in companies like Intercom. It provides a common language and way of thinking about just about every activity in a company.

But to be sure, JTBD is not a silver bullet. It won't cure all of your problems. I believe, however, it has the potential to reframe conversations and change mindsets better than other approaches. Its potential pervasiveness is unmatched, which is why I believe we're seeing stronger interest and take-up of JTBD from a range of disciplines. JTBD isn't just a bolt-on technique, but potentially something that runs through the veins of your organization.

Regardless of your perspective and regardless of affiliation to a particular method, now is the time to learn JTBD and bring it to your work.

Quick Reference: JTBD Plays

The different techniques and approaches covered in this book are not exhaustive. My aim is to provide you with a sample of key practices already existing within the field that cover a range of effort types. From these, you can develop a deeper understanding of jobs theory and its corresponding principles.

JTBD is not a single method—it's a way of seeing. Through the lens of JTBD, you can solve challenges in your organization from big (e.g., corporate growth strategy) to small (e.g., how to structure marketing messages or support articles) across disciplines and roles.

Getting Started

Scope the JTBD domain

You'll have to define the domain for every JTBD effort before beginning. Getting the main job and job performer at the right level and well-formulated is key. Start each JTBD project with this play. Involve the entire team and spend time getting the right JTBD definition.

STEPS:

1. Define the main job.
2. Define the job performer.
3. Form a hypothesis about the job process and circumstances.

EFFORT: *Low to medium.* This step frames your subsequent investigations and ultimately your business. You will need access to key decision-makers to define the primary target. A group of people should be involved. In some cases, the main job and the job performer will be obvious. Other times, you may need to discuss and debate the answer.

WATCH OUT: Getting the right level of abstraction is critical. In doubt, scope the main job broad rather than narrow.

TRY IT: Consider the main job that customers are trying to get done with your solution. If possible, ask a few customers about what they are trying to do. Map out different levels of abstraction by asking "why?" to move up and "how?" to move down. Formulate an answer in a way that is devoid of technology and stable over time. Discuss the main job with colleagues and see if it resonates with them.

SEE:

- Anthony Ulwick. *Jobs to Be Done: Theory to Practice* (Idea Bite Press, 2016)
- Bob Moesta. "Bob Moesta on Jobs-to-be-Done," interview by Des Traynor, *Inside Intercom* (podcast, May 12, 2016)

Discovering Value

1. Conduct jobs interviews

Jobs are not made up or brainstormed in isolation. They are discovered. Prepare to do a lot of interviewing: it's the key way you'll uncover jobs.

EFFORT: *Medium to high.* The level of effort depends on the number of interviews completed. You will need to be able to recruit participants with in-depth sessions. Interviews typically range from 30 minutes to two hours each for 6–20 or more participants. You can do them remotely or in person. Often, multiple interviews contribute to the discovery of jobs, and debriefing sessions are needed. Recording the audio of each session is a good idea, but re-listening to each can take hours.

WATCH OUT: If you've never done qualitative research before, jobs interviews can be overwhelming. The amount of feedback you'll get from just a dozen interviews can be voluminous. Have a good plan for capturing and managing the data you gather. It takes practice to recognize jobs and draw out the key, relevant information you need.

TRY IT: Practice open interviewing techniques with a friend or colleague on a job they are trying to get done, e.g., *prepare a meal*. Record the interview and listen back. Did you avoid yes-or-no questions? Did you let them do most of the talking? Did you dig deeper and say, "Tell me more about that?"

SEE:

- Steve Portigal. *Interviewing Users* (Rosenfeld Media, 2013)
- Giff Constable. *Talking to Humans* (2014)
- Mike Boysen. "A Framework of Questions for Jobs to Be Done Interviews" (*Medium*, 2018)
- Hugh Beyer and Karen Holtzblatt, *Contextual Design* (Morgan Kaufmann, 1998)

2. Run Switch interviews

The Switch method for conducting JTBD interviews is an alternative to jobs interviews. The technique was pioneered by Bob Moesta and Chris Spiek at Re-Wired. Switch interviews are good when you have a known product that you want to improve or create more demand for.

STEPS:

1. Recruit customers.
2. Interview based on the Switch timeline.
3. Analyze the forces of progress.

EFFORT: *Medium to high.* Switch interviews can be done within a few days. Full-blown efforts with a large sample may take weeks. You will need to be able to recruit 6–12 participants and schedule in-depth interviews with them. Each session takes anywhere from 15–60 minutes. Taking notes directly on the Switch timeline eases analysis, which can be done from your raw notes.

WATCH OUT: Switch interviews tend to focus on the product, but the aim is to uncover underlying motivations. It takes practice to keep separate the purchase decision from the job to be done.

TRY IT: Practice elements of the Switch method any time you speak with customers. Start by simply asking, "Tell me about your decision to use the new product." Add more elements of Switch after you get a sense of how customers will respond to that initial question.

SEE:

- Chris Spiek and Bob Moesta. *Jobs-to-Be-Done: The Handbook* (Re-Wired Group, 2014).
- Clayton Christensen et al. "How to Hear What Your Customers Don't Say," Chapter 5 in *Competing Against Luck* (HarperBusiness, 2016)
- Alan Klement. *When Coffee and Kale Compete* (Self-published, 2016)

3. Analyze the Four Forces of Progress

There are four forces that drive behavior of switching from one offering to another. The Four Forces analysis looks at the factors involved in moving from a current state to a new state. A *problem* with an existing solution and the *attraction* of a new one motivate customers to change. But the *uncertainty* about change, along with *habits*, keeps consumers from switching. Ultimately, you're looking for the progress that people want to make when hiring a particular offering.

STEPS:

1. Conduct research.
2. Extract insights around each of the forces.
3. Find your opportunity.

EFFORT: *Medium.* The interviews to research the Four Forces are the most time consuming. You'll need to recruit and schedule interviews with 6–12 participants. After synthesizing the data collected, plan a session with internal stakeholders to review the findings and come up with an action plan.

WATCH OUT: The Four Forces can be approached informally, but actually it requires rigor to complete properly. Don't let the simplicity of the technique suggest that deep research and analysis isn't needed.

TRY IT: Divide a sheet of paper or document into four quadrants and label them with "problems," "attraction," "anxiety," and "habit" going clockwise from the upper left. Then consider your own product or service and what factors are in play for each of the topics. Discuss your findings with a colleague to see if they have the same insights.

SEE:

- Alan Klement. *When Coffee and Kale Compete* (NYC Press, 2016)
- Chris Spiek and Bob Moesta. "Unpacking the Progress Making Forces Diagram," *JTBD Radio* (Feb 2012)

4. Map the main job

A job map visualizes the flow of the main job. This technique was pioneered by Tony Ulwick as part of his Outcome-Driven Innovation (ODI) approach. Job maps can be part of just about every JTBD effort and serve as a cornerstone in JTBD approaches.

EFFORT: *Medium.* An initial job map can be built during the interviewing process and finalized a few days afterward. Iterate on the process and include others for validation.

WATCH OUT: Don't conflate job maps with customer journey maps, service blueprints, or other types of workflow diagrams. The map isn't about your solution, brand, or customers; it's a map of the job to be done.

TRY IT: Take a simple main job, such as *prepare a meal.* Using your own experience, create a hypothesis of the stages based on the universal job map following the rules of formulating job statements. Then speak with at least two meal preparers to understand their process. Does it match your hypothesis? What would you change?

SEE:

- Lance Bettencourt and Anthony Ulwick. "The Customer-Centered Innovation Map," *Harvard Business Review* (May 2008)
- Jim Kalbach. "Experience Maps," Chapter 11 in *Mapping Experiences* (O'Reilly, 2016)
- Tony Ulwick. "Mapping the Job-to-be-Done," *JTBD+ODI blog* (Jan 2017)

Defining Value

1. Find underserved needs

In principle, finding underserved needs is straightforward: identify needs that are important but unsatisfied. In practice, this can be challenging.

STEPS:

1. Gather all desired outcomes.
2. Formulate desired outcome statements.
3. Survey job performers.
4. Find opportunities.

EFFORT: *High*. A full needs analysis following ODI can take months to complete. First, conduct extensive jobs interviews to extract desired outcome statements. Then a survey is needed with a large sample size that can takes weeks to administer. Other alternative approaches are less time-consuming but lack the rigor of ODI.

WATCH OUT: It's difficult to uncover all needs and formulate statements. Getting hundreds of participants to complete a survey with 50 or more items is challenging. While informal approaches exist, cutting corners here generally results in inconclusive or speculative results.

TRY IT: Collect a small set (e.g., a dozen) of need statements for your main job on individual sticky notes. Place each on a whiteboard with a 2 × 2 grid. Prioritize assumed importance and satisfaction. Discuss the opportunities, assumptions, and risks as a group. The result will be invalid, but can sensitize the team to how the principles of prioritizing opportunities work.

SEE:

- Anthony Ulwick. "Turn Customer Input into Innovation," *Harvard Business Review* (Jan 2002)
- Dan Olsen. *The Lean Product Playbook* (Wiley, 2015)
- Scott Anthony et al. *The Innovator's Guide to Growth* (Harvard Business Review Press, 2008)

2. Create goal-based personas

Alan Cooper's goal-driven approach results in personas based on the goals that users have, not demographics.

STEPS:

1. Interview users and distill the key variables in goals.
2. Map interviews to variables.
3. Identify patterns in goals.
4. Describe the resulting persona based on common goals.

EFFORT: *Medium to high.* The primary research to create goal-driven personas can be high. You'll have to recruit interviewees, interview them, and analyze the data. The process can range from days to weeks to complete, depending on the complexity of your situation. Proto-personas are a lightweight alternative based on current knowledge in your team and assumptions.

WATCH OUT: Since personas are summarized based on a hypothetical person, they tend to suggest demographics rather than goals. Your team may misinterpret personas as a reflection of personal traits. Also, personas have a tendency to become irrelevant to your team as projects progress. Work to make them come alive by hanging persona posters around the office or role-playing personas in meetings.

TRY IT: After interviewing job performers, list the key circumstances. Map the interviews to the circumstances and find patterns. Then create jobs-based personas that reflect the different outcomes each has.

SEE:

- Alan Cooper and Robert Reimann. *About Face 2.0: The Essentials of Interaction Design* (Wiley, 2003)
- John Pruitt and Tamara Adlin. *The Persona Lifecycle: Keeping People in Mind Throughout Product Design* (Morgan Kaufmann, 2006).
- Kim Goodwin. *Designing for the Digital Age: How to Create Human-Centered Products and Services* (Wiley, 2009)

3. Compare competing solutions

Use JTBD to compare solutions for a unique view of the competition across product categories. Use needs or process steps as the basis of comparison. To be thorough, survey job performers to find out how they rank each need against the competitors. The intent is to be able to find your opportunity in the competitive landscape from a jobs perspective to differentiate your offering.

STEPS:

1. Determine alternatives to compare.
2. Determine the needs or the job steps to compare.
3. Rank how well each solution meets those needs.
4. Find your sweet spot in the competitive landscape.

EFFORT: *Low to high.* If you've already uncovered needs, getting started with a competitor comparison will be easy. You can estimate how well each need or job step gets done. However, if you need to conduct primary research, the effort will be high. Surveying customers directly on how they compare solutions adds significantly to the effort.

WATCH OUT: If you don't survey job performers, your comparison will be speculative. Nonetheless, you can still discuss the potential differences across solutions as a team to get internal agreement. Be sure to test your assumptions in any event.

TRY IT: Consider all of the ways that people get a main job done related to your solution. Take a small subset of needs or process steps and put them on a table. Select two to three solutions to compare. Discuss as a team how well each one performs relative to the other. Then ask: "Where does your solution fall in the comparison?" "Where might you improve?" "What steps are the most strategic to address?"

SEE:

- Anthony Ulwick. *Jobs to Be Done: Theory to Practice* (Idea Bite Press, 2016)

4. Define a jobs-based value proposition

A value proposition is an explicit definition of the benefit that you will offer to customers. The Value Proposition Canvas (VPC) created by Alexander Osterwalder is a simple tool to help you define the proposition together with your team.

STEPS:

1. Understand the customer profile.
2. Discuss the solution profile.
3. Ensure the fit between the customer and the solution.
4. Form a value proposition statement.

EFFORT: *Low to medium.* The VPC is a tool to foster internal dialogue about your value proposition. Assuming that you have done JTBD research already, it's fairly easy to get started. The difficulty of this effort increases with more stakeholders to include and the complexity of the situation.

WATCH OUT: It's easy to find many examples of how to use the VPC on the web. But many of these resources use imprecise language for jobs and mix problem space and solution space aspects. While the VPC can be an informal tool for internal alignment, it's recommended to put some rigor into the information included, particularly the key jobs to be done that you start with.

TRY IT: Download the VPC online and practice filling it out by yourself for any topic—either your own business or some other existing business. See if you can match the pains and gains to the pain relievers and gain creators.

SEE:

- Alexander Osterwalder, Yves Pigneur, Gregory Bernarda, and Alan Smith. *Value Proposition Design* (Wiley, 2014)

Designing Value

1. Create a development roadmap

Use JTBD to drive the themes in your product roadmap to ensure that your efforts are tied to customer needs. Keep the roadmap at a high level with indefinite delivery dates to provide guidance without committing to specific times.

STEPS:

1. Define the solution direction.
2. Determine the customer needs to pursue.
3. Set a timeline.
4. Align the development effort to the roadmap.

EFFORT: *Medium to high.* Creating an initial format and distribution means may take some effort at first. But after that, extending your roadmap may only take a few hours. A roadmap typically involves input from a range of stakeholders, from key decision-makers to the teams who will implement a solution. Coordinating all of the input adds greatly to the effort.

WATCH OUT: Roadmaps are often confused with detailed project plans, which have defined timelines. Instead, think of roadmaps as painting a broader picture of the sequence of activity. Also, keep in mind that roadmaps change as well. You'll want to update them each quarter or so.

TRY IT: Complete a draft roadmap for development of your solution for the next quarter. See how much of it you can complete in 1–2 hours at first. Use your JTBD framework to align the themes in the roadmap to the sequence of activities you map out.

SEE:

- C. Todd Lombardo, Bruce McCarthy, Evan Ryan, and Michael Connors. *Product Roadmaps Relaunched* (O'Reilly, 2017)

2. Align teams to job stories

Job stories provide a standard format to represent smaller jobs to be done. Derive them from your job map and jobs interviews. Because job stories include information about the circumstances and job steps, they can stand on their own, giving teams the flexibility to bring them in as needed. At the same time, job stories provide confidence that features and capabilities are grounded in fulfilling a need.

STEPS:

1. Understand job stages and circumstances.
2. Formulate job stories.
3. Solve for the job stories.

EFFORT: *Low.* It's fairly easy to generate a set of job stories after you've completed JTBD research, created a job map, and have a roadmap.

WATCH OUT: Job stories have many possible formats and their uses vary. The key is to develop a consistent, stable format that suits your situation. The word "story" implies they may replace user stories in Agile development. But, in most situations, this is not the case: you'll still measure project burn-down rate in Agile development using traditional user stories.

TRY IT: Consider the jobs to be done related to a current effort you are working on. Formulate 6–12 job stories using a common format based on data from your research. Share them for discussion. Then make the stories visible in your team's meetings, workshops, and design sessions to connect your effort to customer needs.

SEE:

- Alan Klement. "Replacing the User Story with the Job Story," *JTBD.info* (Nov 2013); "5 Tips for Writing a Job Story," *JTBD.info* (Nov 2013); and "Designing Features Using Job Stories," *Inside Intercom* (2015)
- Maxim van de Keuken. "Using Job Stories and Jobs-to-be-Done in Software Requirements Engineering," [Thesis, Utrecht University] (Nov 2017)

3. Architect the solution

In software, the structure of a solution, whether a product or service, can be seen independent of the interface with the consumer. You should base the blueprint of a solution on the individual's jobs to be done for longevity and better comprehension of the solution. There are several related existing methods that show how to do this in different situations, such as User Environment Design (UED).

STEPS:

1. Understand users and their work.
2. Identify corresponding focus areas.
3. Model the solution structure.

EFFORT: *Medium to high.* Creating conceptual models of a solution architecture is iterative and must be refined over time. The more complex the solution, the more difficult the challenge will be. Models for extensive software applications or websites may take weeks to research and develop following processes such as UED.

WATCH OUT: Modeling the solution architecture is an abstract endeavor. Others may confuse it for technical architecture on the one side, or interface design on the other. Keep discussions of the underlying model separate from surface-level implementation.

TRY IT: Take a software product, such as a photo-editing program or email client, and try to reverse engineer its underlying architecture. First, list all of the navigation points and functional options and group them into clusters. Then create a simple diagram of the underlying architecture and label the components of the model you derived.

SEE:

- Hugh Beyer and Karen Holtzblatt. *Contextual Design* (Morgan Kaufmann, 1998)
- Indi Young. "Structure Derivation," Chapter 13 in *Mental Models* (Rosenfeld, 2008)

4. Test assumptions

Even with the most thorough JTBD investigation, there is no guarantee that the offering you create will be adopted. Plan to experiment and test the solutions you come up with to ensure a better product-market fit.

STEPS:

1. Formulate hypotheses.
2. Validate or invalidate hypotheses with experiments.
3. Make sense of what you learned and move forward.

EFFORT: *High*. To reduce uncertainty and gain confidence in the direction of your solution, an iterative approach is needed. You may go back to the drawing board many times, requiring a commitment of time and effort to get your product right.

WATCH OUT: In commercial settings it's very challenging to isolate variables in your experimentation. You may get feedback that is potentially misleading or doesn't show direct causality. False positives and false negatives are common.

SEE:

- Travis Lowdermilk and Jessica Rich. *The Customer-Driven Playbook* (O'Reilly, 2017)
- Ash Maurya. *Running Lean* (O'Reilly, 2012)
- Eric Ries. *The Lean Startup* (Crown, 2011)
- Steve Blank. *Four Steps to the Epiphany*, 2nd Ed (K & S Ranch, 2013)
- Michael Schrage. *The Innovator's Hypothesis* (MIT Press, 2014)

Delivering Value

1. Map the consumption journey

Journey maps show the interaction that customers have with your brand or offering. The intent of a journey map is to illustrate the consumption jobs that customers have. It is fundamentally different than a job map, which shows what a job performer is trying to get done independent of a given solution. Journey maps provide valuable insight into how people will find, acquire, and use your solution.

STEPS:

1. Initiate a journey-mapping project.
2. Investigate the steps in consumption.
3. Illustrate the journey in a diagram.
4. Align around the consumption journey.

EFFORT: *Medium to high.* If you have existing research on which to base a journey map, the effort may be fairly straightforward. Conducting primary research adds to the overall time and resources needed.

WATCH OUT: Be sure to keep the job map separate from the journey map. There's an important distinction between the two: job maps diagram the steps in getting the main job done; a journey shows the relationship of a customer to a company's offering. Also be sure to include others in the process, making sense of the journey map together. It is a means to an end: to foster conversations within your team.

SEE:

- Lance Bettencourt. *Service Innovation* (McGraw Hill, 2010)
- Anthony Ulwick and Frank Grillo. "Can Bricks and Mortar Compete with On-line Retailing?" *Whitepaper* (2016)
- Jim Kalbach. "Customer Journey Maps," Ch 4 in *Mapping Experiences* (O'Reilly, 2016)

2. Onboard customers successfully

Onboarding customers into a service is critical to their long-term success. JTBD provides a consistent way to see how the objective that customers are trying to achieve is independent of your solution. This insight can help guide onboarding efforts. You ultimately want to onboard the customer not only into your product, but also into the job.

STEPS:

1. Learn about the customers and segment them into different learning types.
2. Determine the optimal sequence of tasks for each type.
3. Design the onboarding experience.

EFFORT: *Medium to high.* There are many touchpoints in a typical onboarding process and lots of details to consider. Iterating and refining an onboarding program can take considerable effort and time.

WATCH OUT: Different types of customers will have different onboarding needs. It may be tricky to segment users and provide the most relevant onboarding experience for them.

TRY IT: Take your current onboarding flow and compare how it addresses the different learning types outlined in this play. Consider ways to improve the flow once you've completed your analysis.

SEE:

- Alan Klement. "Design for Switching: Create Better Onboarding Experiences." *JTBD News* (2014)
- Samuel Hulick. "Applying Jobs-to-Be-Done to User Onboarding, with Ryan Singer!" *UserOnboard* (2016)

3. Maximize customer retention

We live in an increasingly subscription-based economy. With recurring revenue at stake, maintaining ongoing relationships with customers is critical to business success. Finding the first thought of cancelling can shed light on what you can do to prevent churn.

STEPS:

1. Conduct cancellation interviews using the Switch technique.
2. Find patterns in cancellation reasons.
3. Address the root causes to prevent churn and increase retention.

EFFORT: *Medium.* You can gain considerable insight with just a handful of cancellation interviews, but the more, the better. Reviewing the data will take some effort in any event.

WATCH OUT: Recruiting churned customers may be difficult and require an incentive. In some cases, it may be hard for interview participants to reconstruct their reasons for cancelling, particularly in B2B situations where multiple decision-makers may be involved.

TRY IT: Find a friend or colleague who has recently canceled a subscription service of any kind. Use the Switch technique of interviewing to reconstruct their decision-making process and find the first thought. Reflect together on that moment and consider the job to be done. Then ask, "What could the service provider have done to prevent the cancellation?"

SEE:

- Ruben Gamez. "Doing SaaS Cancellation Interviews (the Jobs-to-be-Done Way)" *ExtendsLogic* (2015)

4. Provide relevant support

In support situations, people don't necessarily ask for what they want directly. For one, they may use the wrong language. But they may also have already devised a solution to their problem in their mind and ask about getting the wrong job done. Agents have to clarify the real need before trying to fix a customer issue. JTBD can help get to the best resolution.

STEPS:

1. Listen for the job.
2. Clarify and assess as to the customer's intent.
3. Resolve the issue.

EFFORT: *Low.* It's fairly simple to instruct support agents to probe for the job. Active and empathetic listing is involved.

WATCH OUT: At the micro-job level, thinking about the job tends to blend with the solution. It's hard to separate the two, but possible. It will take some effort to get support agents trained in thinking in terms of jobs and not product tasks.

TRY IT: Get a sample of previous support requests and try to identify the job to be done in each. Formulate your job statements using the rules of JTBD outlined in Chapter 2, "Core Concepts of JTBD."

(Re)developing Value

1. Survive disruption

According to Clayton Christensen, the father of JTBD theory, identifying the jobs that people are trying to get done stands at the core of understanding market disruption. People will switch to solutions that get the job done simpler, quicker, and cheaper. Together with Maxwell Wessel, he offers a lightweight diagnostic model based on JTBD to find where potential disruption may originate by comparing the strengths and weaknesses of competitors in terms of getting the job done.

STEPS:

1. Determine the strengths of the disruptor.
2. Identify your own company's relative advantages.
3. Evaluate barriers.

EFFORT: *Low.* A basic analysis using this technique can be done in a session or two with a limited set of stakeholders. The key is to have the right decision-makers together for the discussion. Even with deeper investigation and data to include, this technique has a relatively low effort.

WATCH OUT: Although there is potentially a high-level impact, the particular approach outlined by Wessel and Christensen lacks rigor. You'll need to have key decision-makers at an organization involved for agreement. It doesn't seek to be exhaustive, so there is inherent prioritization based on strategic imperatives and company vision.

SEE:

- Maxwell Wessel and Clayton Christensen. "Surviving Disruption" (*Harvard Business Review*, 2012)

2. Create a jobs-based strategy

The JTBD perspective offers a new way of looking at strategy from the customer's perspective, from the outside-in. Focusing on the job allows organizations to maintain a constant strategic imperative—get the customer's job done—even as technology changes.

Developed by Tony Ulwick and his company Strategyn, the growth strategy matrix presents a typology of different jobs-based strategies. With the knowledge of products and services that get the job done cheaper and quicker, organizations can use JTBD to achieve more predictable growth. Offerings that get a job done better than others win in the marketplace.

STEPS:

1. Segment customers by JTBD.
2. Decide on the type of strategy to follow.
3. Determine solutions needed to get the job done.
4. Craft a value proposition and create marketing campaigns.

EFFORT: *High*. Creating a jobs-based strategy requires rigorous upfront research. You'll need to ensure that you have a complete set of input from job performers across the market. Decision-making for this technique happens at the highest level of an organization, so getting the right stakeholders on board may be challenging, but critical.

WATCH OUT: Deriving a strategy based on JTBD requires a high degree of precision and solid supporting data. Otherwise, you may be making decision that impact your entire business based on assumptions.

TRY IT: Consider your current strategy and plot it on the growth strategy matrix. What are the implications of that strategy type to your operations? How might you improve or refine the strategy based on the position in the matrix?

SEE:

- Anthony Ulwick. "The Jobs-to-be-Done Growth Strategy Matrix" (*JTBD Blog*, 2017)

3. Organize around jobs

Reorgs tend to focus overly much on hierarchy and lines of reporting. As a result, there is often little difference in the net effect of a reorg. As an alternative, try making JTBD one of your organizing dimensions. This puts the focus directly on customer-centered thinking in a way that is inherent to the company structure.

STEPS:

1. Cluster jobs into local groupings.
2. Organize based on jobs.
3. Set success metrics and measurements.

EFFORT: *Medium to high.* Adjusting how a company is organized is not an easy task. You'll also have to spend considerable energy explaining the JTBD approach to organization to your colleagues and convincing them of the value. To get started quicker, though, you can create tertiary organizational structures within an existing hierarchy, such as working groups, guilds, or chapters, that focus on a given job.

WATCH OUT: Organizing around jobs will likely represent a new way of thinking about your work and your business. Blurring lines of reporting and changing communication patterns may disorient teams initially.

TRY IT: On paper, consider how you might reorganize your company or, more simply, your department, based on JTBD. List all of the existing functions and compare them to stages in getting the main job done. What would the implications be for the org chart? What is gained and what is lost? How would work or communication be different for the teams involved? See if there is a logical matrix-like structure to introduce jobs thinking into the org chart.

SEE:

- Clayton Christensen et al. "Integrating Around a Job," Ch 7 in *Competing Against Luck* (HarperBusiness 2016)

4. Consider ways to expand market opportunities

A jobs-based approach can help you expand your view of the market. The way to do so is to look broader at getting more jobs done or consider the next level up in the jobs hierarchy of goals. In other words, focus on the progress that people are trying to make in their lives as they seek solutions to fulfill their needs.

STEPS:

1. Look at the progress that people want to make.
2. Ask, "What business are we really in?"
3. Reframe your offering.

EFFORT: *Low.* It's fairly easy to speculate on how you might expand. However, acting on concepts you come up with may be existential for a company and extremely difficult to implement.

WATCH OUT: There is no right or wrong answer to reflecting on how to expand your business. You'll have to negotiate responses from your team to come to an agreement. In some cases, there may be multiple, viable answers. In the end, it's the discussion that's important and getting others to open their minds to new possibilities.

SEE:

- Clayton Christensen, Scott Cook and Taddy Hall. "Marketing Malpractice: The Cause and the Cure," *Harvard Business Review* (Dec 2005)

Resources on JTBD

Adams, Paul. "The Dribbblisation of Design," *Inside Intercom* (Sep 2013)

Anthony, Scott, Mark Johnson, Joseph Sinfield, and Elizabeth Altman. *The Innovator's Guide to Growth* (Harvard Business Press, 2008)

Bates, Sandra. *The Social Innovation Imperative* (McGraw-Hill, 2012)

Berkun, Scott. *The Myths of Innovation* (O'Reilly, 2007)

Berstell, Gerald and Denise Nitterhouse. "Looking 'Outside the Box': Customer Cases Help Researchers Predict the Unpredictable," *Marketing Research* (1997)

Bettencourt, Lance. *Service Innovation* (McGraw-Hill, 2010)

Bettencourt, Lance and Anthony Ulwick. "The Customer-Centered Innovation Map," *Harvard Business Review* (May 2008)

Beyer, Hugh and Karen Holtzblatt. *Contextual Design* (Morgan Kaufmann, 1998)

Blank, Steve. *Four Steps to the Epiphany* (2005)

Blank, Steve. "An MVP Is Not a Cheaper Product, It's About Smart Learning," [blog] (Jul 2013)

Blenko, Marcia, Michael Mankins, and Paul Rogers. *Decide & Deliver* (Bain & Company, 2010) https://www.bain.com/insights/decide-and-deliver/

Boysen, Mike. "A New Look at the Buyer Journey—as a Consumption Chain Job-to-be-Done," *Medium* (November, 2016) https://mikeboysen.com/a-new-look-at-the-buyer-journey-as-a-consumption-chain-job-to-be-done-fde28a6fa98d

Boysen, Mike. "If You Can't Identify an Exit Strategy, You Can't Identify Your Market," *Medium* (May 2017) https://mikeboysen.com/if-you-cant-identify-an-exit-strategy-you-can-t-identify-your-market-jtbd-d39961539618

Boysen, Mike. "What #JobsToBeDone Is, and Is Not," *Medium* (Dec 2017) https://medium.com/@mikeboysen/what-is-a-job-3eb1e65c7810

Boysen, Mike. "How to Get Results for Jobs to Be Done Interviews," *Medium* (Mar 2018) https://medium.com/@mikeboysen/jobs-to-be-done-interviews-79623d99b3e5

Caldicott, Sarah Miller. "Ideas-First or Needs-First: What Would Edison Say?" [white paper] (Strategyn, 2009) https://strategyn.com/files/ideas-first-or-needs-first-what-would-edison-say-strategyn/ideas-first-or-needs-first-what-would-edison-say-strategyn.pdf

Christensen, Clayton. *The Innovator's Dilemma* (Harvard Business Press, 1997)

Christensen, Clayton. *The Innovator's Solution* (Harvard Business School Press, 2003)

Christensen, Clayton, Scott Anthony, Gerald Berstell, and Denise Nitterhouse. "Finding the Right Job for Your Product," *MIT Sloan Management Review* (Apr 2007) https://sloanreview.mit.edu/article/finding-the-right-job-for-your-product

Christensen, Clayton, Scott Cook, and Taddy Hall. "Marketing Malpractice: The Cause and the Cure," *Harvard Business Review* (Dec 2005) https://hbr.org/2005/12/marketing-malpractice-the-cause-and-the-cure

Christensen, Clayton, Taddy Hall, Karen Dillon, and David S. Duncan. *Competing Against Luck* (HarperBusiness, 2016)

Christensen, Clayton, Taddy Hall, Karen Dillon, and David S. Duncan. "Know Your Customers' 'Jobs to Be Done'," *Harvard Business Review* (Sep 2016)

Christensen, Clayton and Laura Day. "Integrating Around the Job to Be Done" [white paper] Harvard Business School (May, 2016) https://hbr.org/product/integrating-around-the-job-to-be-done-module-note/611004-PDF-ENG

Constable, Giff. *Talking to Humans* (2014)

Cooper, Alan and Robert Reimann. *About Face 2.0: The Essentials of Interaction Design* (Wiley, 2003)

Cooper, Alan. *The Inmates Are Running the Asylum* (SAMS, 1999)

Deloitte. "Customer-Centricity: Embedding It into Your Organisation's DNA" [white paper] (2014) https://www2.deloitte.com/content/dam/Deloitte/ie/Documents/Strategy/2014_customer_centricity_deloitte_ireland.pdf

Drucker, Peter. *The Practice of Management* (Harper & Brothers, 1954)

Drucker, Peter. *Innovation and Entrepreneurship* (Harper & Row, 1985)

Fenty, Karin. "The Business Impact of Investing in Experience," [report] (Apr 2018)

Fogg, B. J. "Fogg Behavior Model," *behaviormodel.org* (2019)

Gamez, Ruben. "Doing SaaS Cancellation Interviews (the Jobs-to-be-Done Way)," *ExtendsLogic* (Oct 2015) http://www.extendslogic.com/business/jobs-to-be-done-cancel-interviews/

Garret, Jesse James. *Elements of User Experience* (New Riders, 2002)

Gibbons, Sarah. "User Need Statements: The 'Define' Stage in Design Thinking," *Nielsen Norman Group* (Mar 2019)

Goodwin, Kim. *Designing for the Digital Age: How to Create Human-Centered Products and Services* (Wiley, 2009)

Gothelf, Jeff. *Lean UX* (O'Reilly, 2013)

Hackos, JoAnn and Janice C. Redish. *User and Task Analysis for Interface Design* (Wiley, 1998)

Hulick, Samuel. "Applying Jobs-to-Be-Done to User Onboarding, with Ryan Singer!" *UserOnboard* (2017) https://www.useronboard.com/ryan-singer-user-onboarding-jtbd/

IBM. "Needs Statements," *IBM Enterprise Design Thinking Toolkit* (Aug 2018) https://www.ibm.com/design/thinking/page/toolkit/activity/needs-statements

Joffrion, Emily Fields. "The Designer Who Changed Airbnb's Entire Strategy," *Forbes* (Jul 2018)

Johnson-Laird, Philip. *Mental Models* (Harvard University Press, 1983)

Keuken, Maxim van de. "Using Job Stories and Jobs-to-be-Done in Software Requirements Engineering," [Thesis, Utrecht University] (Nov 2017)

Klement, Alan. "5 Tips for Writing Job Stories," *JTBD.info* (Nov 2013). https://jtbd.info/5-tips-for-writing-a-job-story-7c9092911fc9

Klement, Alan. "Replacing the User Story with the Job Story," *JTBD.info* (Nov 2013)

Klement, Alan. "Designing Features Using Job Stories," *Inside Intercom* (Dec 2013) https://blog.intercom.com/using-job-stories-design-features-ui-ux/

Klement, Alan. "Design for Switching: Create Better Onboarding Experiences," *JTBD News* (Jul 2014) http://jobstobedone.org/news/design-for-switching-create-better-onboarding-experiences/

Klement, Alan. *When Coffee and Kale Compete* (NYC Press, 2016)

Klement, Alan. "The Forces of Progress," *JTBD.info* (May 2017) https://jtbd.info/the-forces-of-progress-4408bf995153

Kniberg, Henrik and Anders Ivarsson. "Scaling Agile @ Spotify with Tribes, Squads, Chapters & Guilds" [white paper] (Oct 2012) https://blog.crisp.se/wp-content/uploads/2012/11/SpotifyScaling.pdf

Kupillas, Kevin C. "May the Forces Diagram Be with You, Always," *JTBD.info* (Sep 2017) https://jtbd.info/may-the-forces-diagram-be-with-you-always-applying-jtbd-everywhere-b1b325b50df3

Levitt, Theodore. "Marketing Myopia," *Harvard Business Review* (Jul 1960)

Lombardo, C. Todd, Bruce McCarthy, Evan Ryan, and Michael Connors. *Product Roadmaps Relaunched* (O'Reilly, 2017)

Lowdermilk, Travis and Jessica Rich. *The Customer-Driven Playbook* (O'Reilly, 2017)

Maurya, Ash. *Running Lean* (O'Reilly, 2012)

McGrath, Rita Gunther. *The End of Competitive Advantage* (Harvard Business Review Press, 2013)

Minor, Dylan, Paul Brook, and Josh Bernoff. "Are Innovative Companies More Profitable?" *MIT Sloan Management Review* (2017) https://sloanreview.mit.edu/article/are-innovative-companies-more-profitable/

Murphy, Lincoln. "Understanding Your Customer's Desired Outcome," *customer-centric growth by lincoln murphy* (no date) https://sixteenventures.com/customer-success-desired-outcome

Norman, Don. (2009) "Technology First, Needs Last," jnd.org (2009)

Olsen, Dan. *The Lean Product Playbook* (Wiley, 2015)

Osterwalder, Alex and Yves Pigneur. *Business Model Generation* (Wiley, 2010)

Osterwalder, Alexander, Yves Pigneur, Gregory Bernarda, and Alan Smith. *Value Proposition Design* (Wiley, 2014)

Porter, Michael and Mark R. Kramer. "Creating Shared Value," *Harvard Business Review* (Jan–Feb 2011)

Portigal, Steve. *Interviewing Users* (Rosenfeld Media, 2013)

Reeves, Martin, Knut Haanaes, and Janmejaya Sinha. *Your Strategy Needs a Strategy* (Harvard Business Review Press, 2015)

Rhea, Brian. "Customer Acquisition & Customer Retention," [blog] (Mar 2019)

Ries, Eric. *The Lean Startup* (Crown, 2011)

Rogers, Everett. *Diffusion of Innovations*, 5th ed. (Simon and Schuster, 2003)

Satell, Greg. "The 4 Things You Need to Know to Make Any Business Successful," *DigitalTonto* (Jan 2013) http://www.digitaltonto.com/2013/the-4-things-you-need-to-know-to-make-any-business-successful/

Schrage, Michael. *The Innovator's Hypothesis* (MIT Press, 2014)

Slocum, David. "Can We Get Beyond Customer Centricity?" *Forbes* (April 2017)

Spiek, Chris and Bob Moesta. "Unpacking the Progress Making Forces Diagram," *JTBD Radio* (Feb 2012)

Spiek, Chris and Bob Moesta. *Jobs-to-Be-Done: The Handbook* (Re-Wired Group, 2014)

Traynor, Des. "Bob Moesta on Jobs-to-Be-Done," *Inside Intercom* (May 2016) https://www.intercom.com/blog/podcasts/podcast-bob-moesta-on-jobs-to-be-done/

Traynor, Des. Intercom. *Intercom on Jobs-to-be-Done* (2016)

Ulwick, Anthony. "Turn Customer Input into Innovation," *Harvard Business Review* (2002)

Ulwick, Anthony. *What Customers Want* (McGraw Hill, 2005)

Ulwick, Anthony. *Jobs to Be Done: Theory to Practice* (Idea Bite Press, 2016)

Ulwick, Tony. "The Jobs-to-Be-Done Growth Strategy Matrix," *JTBD+ODI Blog* (Jan 2017) https://jobs-to-be-done.com/the-jobs-to-be-done-growth-strategy-matrix-426e3d5ff86e

Ulwick, Tony. "Mapping the Job-to-Be-Done, " *JTBD+ODI blog* (Jan 2017)

Ulwick, Anthony. "What Is Jobs-to-Be-Done?" *JTBD+ODI blog* (Feb 2017) https://jobs-to-be-done.com/what-is-jobs-to-be-done-fea59c8e39eb

Ulwick, Tony and Frank Grillo. "Can Bricks and Mortar Compete with On-Line Retailing?" [white paper] (2016) http://resources.hartehanks.com/guides/can-bricks-and-mortar-compete-with-on-line-retailing

Ulwick, Anthony and Lance Bettencourt. "Giving Customers a Fair Hearing," *MIT Sloan Management Review* (Apr 2008) https://sloanreview.mit.edu/article/giving-customers-a-fair-hearing/

Wessel, Maxwell and Clayton Christensen. "Surviving Disruption," *Harvard Business Review* (Dec 2012)

Wilson, Mark. "Trulia Is Building the Netflix for Neighborhoods," *Fast Company* (Aug 2018)

Wunker, Stephen, Jessica Wattman, and David Farber, *Jobs to Be Done: A Roadmap for Customer-Centered Innovation* (AMACON, 2016)

Young, Indi. *Mental Models* (Rosenfeld Media, 2008)

Index

C

D

E

F

G

H

I

J

K

O

P

T

U

V

W

Y

Z

Acknowledgments

Writing is anything but solitary. This book is the result of my collaboration with dozens of people. I'm highly appreciative for all of the input, feedback, and support I've gotten for this project over the past year.

First, I'd like to thank the numerous reviewers who graciously provided direct feedback along the way. In particular, Andrea Hill and Dale Halvorson served as formal technical reviewers for the project. Andrea and I worked together at LexisNexis, and I've learned a lot from her over the years. Dale's insights into JTBD have been extremely helpful to the creation of the book. Thanks to both!

I'm indebted to the many others who have read and reviewed early drafts of this book. Your contributions were invaluable to me, including: Matthias Feit, Brent Schmidt, Brian Rhea, Steph Troeth, Tony Ulwick, Sam Dix, Andrei Radulescu, Sian Townsend, Michael Schrage, Kirsten Mann, Leo Frishberg, Jonathan Horowitz, Liz Trudeau, Eckhart Böhme, and Laura Klein.

A special thank-you goes to my friends and colleagues who contributed case studies and examples: Jake Mitchell, Vito Loconte, Agustin Soler, Steph Troeth, and Kathryn Papadopoulos.

And of course, thank you to Marta Justak, Lou Rosenfeld, and everyone else at Rosenfeld Media for making this project possible. As well, thanks to those who provided testimonials for the book: Tony Ulwick, Jeff Gothelf, Richard Dalton, Des Traynor, Melissa Perri, Lada Gorlenko, Giff Constable, Peter Merholz, Andrea Gallagher, and Bob Moesta.

I'd also like to show appreciation for those who gave me permission to reprint and use their materials: Strategyzer (www.strategyzer.com) for the Value Proposition Canvas; the authors of *Product Roadmaps Relaunched* for the example roadmap; Martin Ramsin, co-founder and CEO of CareerFoundry, for permission to use a photo of a mapping session at the CareerFoundry; BJ Fogg for permission to use the Fogg Behavior Model diagram; and Malia Basche, Hilary Dwyer, and Leslie Pearlson at LogMeIn for permission to use the GoToWebinar case story.

Finally, this book is a tribute to the pioneers and thought leaders in the field of JTBD, starting with Clayton Christensen, Bob Moesta, and everyone at the Re-Wired group who have led the way with their groundbreaking work. Tony Ulwick, Mike Boysen, and others at Strategyn have been particularly influential to my understanding of JTBD. And thanks to Des Traynor and the good folks at Intercom for showing us all practical uses of JTBD.

Thank you all!

About the Author

JIM KALBACH is a noted author, speaker, and instructor in user experience design, information architecture, and strategy. He is currently Head of Customer Experience at MURAL, the leading online whiteboard. Jim has worked with large companies, such as eBay, Audi, Sony, Elsevier Science, LexisNexis, and Citrix.

In 2007, Jim published his first full-length book, *Designing Web Navigation* (O'Reilly, 2007). His second book, *Mapping Experiences* (O'Reilly, 2016), focuses on the role of visualizations in strategy and innovation. He blogs at experiencinginformation.com and tweets under @jimkalbach.